Being a Biomedical Entrepreneur

Growth of the Biomedical Industry

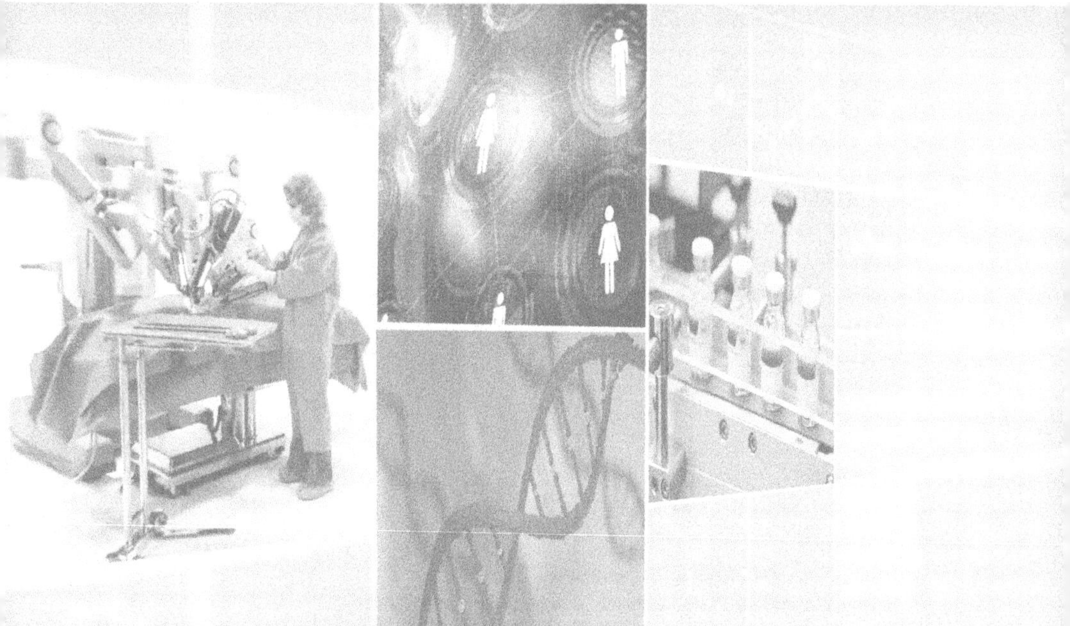

Being a Biomedical Entrepreneur

Growth of the Biomedical Industry

Jen-Shih Lee

Global Monitors Inc. and University of Virginia, USA

World Scientific

NEW JERSEY • LONDON • SINGAPORE • BEIJING • SHANGHAI • HONG KONG • TAIPEI • CHENNAI • TOKYO

Published by

World Scientific Publishing Co. Pte. Ltd.
5 Toh Tuck Link, Singapore 596224
USA office: 27 Warren Street, Suite 401-402, Hackensack, NJ 07601
UK office: 57 Shelton Street, Covent Garden, London WC2H 9HE

British Library Cataloguing-in-Publication Data
A catalogue record for this book is available from the British Library.

BEING A BIOMEDICAL ENTREPRENEUR
Growth of the Biomedical Industry

ISBN 978-981-3270-42-8

For any available supplementary material, please visit
https://www.worldscientific.com/worldscibooks/10.1142/10991#t=suppl

This work is dedicated to the biomedical
entrepreneurs, engineers, technologists, scientists,
mathematicians and physicians
who build the modern biomedical industry and
give a full life to the people of the world

Preface

In 2015, US healthcare spending increased 6% to reach $3.2 trillion, $9,990 per person, or 18% of US GDP. The spiraling increase in healthcare costs may be one reason that some 12% of Americans were uninsured in the second quarter of 2017. In 2016, the market size for US pharmaceutic products is $446 billion while that for US medical devices is $133 billion. China's healthcare and biomedical industry continue to develop at an astonishing rate, spending is projected to grow from $357 billion in 2011 to $1 trillion in 2020. With the completion of the International Human Genome Project in 2003, scientists, engineers and physicians are developing personalized medicine to better predict and prevent disease, to improve medicine and to make accurate diagnosis.

On the other hand, obesity, diabetes, chronic pulmonary obstructive disease, end stage renal disease, physical disabilities and deadly infectious diseases are just too prevalent throughout the world. Even though we have gained a greater understanding of the underlying biology of many cancers and are creating new avenues in screening and early detection, effective screening tests for early detection do not exist for many cancers.

These social, economic and scientific challenges need to be addressed by the professionals in the biomedical industry and biomedical entrepreneurs. They are bound to create enterprises that can innovate and improve current diagnostics and treatments of diseases, to increase their safety and efficacy and to deliver higher quality healthcare at an affordable cost. This assertion calls for building-up biomedical entrepreneurship throughout the world. This book is written to help make your effort in being a biomedical entrepreneur a smooth sailing and a high chance for success.

To achieve diversity, which is essential for BME students, many BME education programs in the US have already overwhelmed their undergraduates with courses in engineering, biology and medicine. A student survey of career choices indicates that BME undergraduates want to gain business and entrepreneurial experience so that, in ten to twenty years following university graduation, they can be entrepreneurs, their top career choice, building their enterprises in medical technology that can resolve or alleviate the medical problems of their friends, relatives and patients.

Judging from the enthusiasm of, and the extra effort spent by, students at the University of Virginia and University of California, San Diego, in taking the "BME Entrepreneurship" course, the author believes that many of them have already taken a giant step toward their entrepreneurial careers. They have spent a great deal of effort to work out a business plan for the course that tells them how qualified they are as entrepreneurs, how critical the need to resolve the medical problem is, how their product uniquely resolves the problem and how their product can generate profits for their company. Support for these assertions has been gathered from interviews by these students with many medical and engineering experts.

This book has been written for these students and others who want to become biomedical entrepreneurs. The first chapter introduces the biomedical industry and describes its growth in the last eighty years. The second chapter describes the establishment of biomedical engineering departments in many research universities and what you need to learn and work on to become a biomedical engineer.

The next sixteen chapters were written to address the following five questions and related issues:

- What is required to become a biomedical entrepreneur?
- How can innovation or improvement be demonstrated to benefit patients while generating profits for the company?
- Who are the biomedical entrepreneurs whose engineering and business ingenuity led them to develop successful enterprises?
- What are the unique challenges and opportunities that biomedical enterprises will face?
- How do you globalize your enterprise and improve your invention for the delivery of affordable healthcare to the Third World?

In this edition, we expand significantly the coverage of the biomedical industry to new technologies. These include tissue engineering, genetic editing and sequencing technology, medical imaging, new radiation therapy, wearable devices, 3-D printing and manufacturing, and organs on a chip, robotics and artificial intelligence, and digital health and information technology. Companies that have been built on these technologies have been added throughout the book to show the challenge in becoming successful and profitable.

The book concludes with two chapters on how to invest, after your enterprise has achieved success, in you, your company, society, the nation, the people and the world, and how to succeed in biomedical engineering entrepreneurism with really trying (by Shu Chien).

To all of you who want to become entrepreneurs for the worldwide betterment of healthcare, the author extends his best wishes. It is his hope that you can be a contributor in making diagnostics and therapies safer, more effective, and more affordable.

Jen-Shih Lee

Acknowledgments

My previous book is written about ***Biomedical Engineering (BME) Entrepreneurship***. The writing of the book and its new version ***Being a Biomedical Entrepreneur*** was made possible by the enthusiastic support of and valuable critiques from many contributors, who are identified below.

First, a course with the same title was taught at the University of Virginia (UVA) in 2003. The author is in debt to the stimulating lectures delivered by Dr. Howard Diamond, CEO, Diamond Electro-Tech, Inc.; Dr. Wendell Dunn, Professor of Entrepreneurship, Darden School of Graduate Business, UVA; Dr. George Gilles, Professor of Mechanical and BME, UVA; Dr. Robert Munzner, Former Chief, of the Neurological Device Branch, FDA; Dr. Henry Reeves, Director, JMU Small Business Development Center, SBA; Mr. Sheridan Snyder, Chairman and CEO, Upstate Biotechnology; and Dr. T. Woodworth, Virginia's CIT. A significant portion of this book has been constructed from the lecture notes of these invited speakers.

In 2007, the course was shortened and taught at UCSD. Several lectures were delivered by Mr. Eric Eng, CPA, CEA, LLP; Mr. Erik Engleson, Former CEO, Target Therapeutics; Dr. Michael Heller, Professor of Bioengineering, UCSD; Dr. Duane Roth, CEO, CONNECT; Dr. John Watson, President of AIMBE (2007-2008); Dr. Jim Wang, CEO, Harmonic Tech; and Mr. Paul Wang, Senior VP of Marketing and Engineering, Spectragraphics Corp.

The author wants to express his thanks to the students of both classes for making numerous comments in class. They interviewed many experts in the field of their company developments, spent a great deal of effort to prepare their business plans and enthusiastically presented the vision, strategy and actions necessary to develop their companies.

To learn more about the legendary Mr. U. A. Whitaker, the author was honored to interview Drs. Burtt Holmes and Ruth Whitaker Holmes at The Whitaker Foundation in Arlington, Va. At the 24th NAPSE

Annual Meeting in DC, the author interviewed Earl Bakken to seek his advice on writing this book. Later, the author interviewed him once more at his Big Island home in Hawaii. When asked how students should prepare themselves for a career in BME, he replied unequivocally, "Learn to speak Chinese because of the prominent role of China in global commerce." In this Hawaii visit, the author had an opportunity to see Bakken's "integrated healing center' in Waimea, Hawaii, which has turned Hawaii's Big Island into a "Healing Island" that befits the people who live or visit there. The author appreciates greatly the opportunity to interview Mr. Robert Capon and Dr. Joe Linden on the building-up of Adenosine Therapeutics.

As a graduate of the CONNECT's Springboard Program, the author is indebted to the entrepreneurial coaching of Messrs. Marty Turock, John Plavan, Ruprecht von Buttlar and Charles Novitsky. He is grateful for the coaching effort of Messrs. Les Briney and Michael Lutz when Global Monitors, Inc. (GMI) participated in the 2008 Tech Coast Angels' Quick Pitch Competition.

Special thanks are due to Drs. Charles Wang, CEO and Lily Wang, President, Optodyne. Their success in patenting, manufacturing and marketing their Laser Doppler Displacement Measuring Systems serves as our role model in entrepreneurship. They are the ones who convinced the author to proceed with the building-up of GMI. Mr. Paul Wang had spent many hours in helping the author write the business plan of Global Monitors, Inc. The author learned a great deal from his entrepreneurial experiences as he had worked at various senior management positions in engineering, marketing and sales for seven large to small start-up companies. Paul has provided many constructive critiques of Chapter 12, "Writing and Presenting the Business Plan".

Dr. Chor Tan, initiator of the Engineering Management Certificate International of ASME and former Dean of Engineering, Cooper Union University, provided excellent suggestions for improving the presentations on engineering management in the chapters of Part IV, Building-up the Enterprise. Thanks also go to Dr. Jackson S. N. Tung, Chief Scientist, Yunnan Health and Development Research Association and Dr. S. X. Cai, Professor of Biomechanics, Chongqing University for the improvements of the chapter on globalization. The author received considerable help from Dr. Shyhhau Wang, National Cheng-Kung University on outsourcing to and doing business in Taiwan. I also thank

Mr. Kenneth Weitzman for his input on patenting and Mr. Dennis White for his contributions to outsourcing, hardware manufacturing and ventilator design.

To the author's dear BME colleagues Drs. Peter Katona, Michael Khoo, John Linehan, and Robert Nerem, the author is grateful for their excerpts that were specifically written for Chapter 19 and Dr. Shu Chien who wrote the stimulating and challenging Chapter 20 on BME Entrepreneurism.

Thanks go to my friends Drs. Bill Chen, Peter Chen, Philip Chiu, Wilfred Huang, Alvin Tong and Richard Wang who helped to edit several chapters of this book. Mr. Albert Lee, VP of Production of GMI, made many valuable contributions to clearly present the steps that are involved in entrepreneurship in this book. More recently, thanks are due to Dr. Warren Huang, former CEO of Afton Chemical and Mr. Dennis Bourque, CEO of ReMend, on the challenges to becoming a successful entrepreneur.

The author appreciates the help from Professors Shu Chien and Peter Chen of University of California San Diego, Professors Shasa Popel and Art Shoukas of Johns Hopkins University, and Professors Jean-Michael Maarek and Jesse Yen of University of Southern California in conducting the career survey with their students.

In 2002, the author had the opportunity to work with a business planning team for the "Venture Assessment" course offered by Dunn at UVA. The students of Darden School of Graduate Business are Audrey Cao, Nirav Desai and Nitin Jain, who chose CardioMonitors, Inc. as their company for the due diligence study. In just one semester and with regular weekly meetings, they conducted a thorough due diligence analysis of the business aspects of a blood volume monitor. This was the author's first ever experience in working with a group of exceptional business students and seeing their business-like pursuit of due diligence. This experience convinced the author that the success of an entrepreneurial team can only be achieved by people who are knowledgeable in technology development and business operation working together with the least number of flaws and best collaboration.

The author wants to express his thanks to websites such as Google, Wikipedia and those on current news and financial analyses.

The author would like to acknowledge the help and advice given by Ms. Chelsea Chin and Dr. Yan Hong Ng of World Scientific Publishing on writing the book. To finalize the book for publication, the fantastic editing of the entire book by Mr. George Robinson of WritingEnglish.com and Kim Tan and Yolande Koh of World Scientific Publishing are deeply appreciated.

The author wishes to advise that he has no financial interest in the companies that are mentioned in this book.

This book is also dedicated to my grandchildren Angelina, Zoe, Justin, Max, Annabelle and Trinity with the wish that some of them will become a biomedical entrepreneur, scientist, engineer or physician.

Finally the author acknowledges the support and help from his wife Dr. Lain-pin Lee in building a medical device company and writing this book to promote the development of biomedical entrepreneurship.

Contents

Chapter 1

The Biomedical Industry

1.1. Challenges and Opportunities

The biomedical industry is a knowledge-intensive industry that is constantly generating, and adapting to, new technologies. It is characterized by innovation, ingenuity and a constant search for new challenges. The biomedical industry can be divided into four sectors:

1. ***Pharmaceutical Industry***
2. ***Biotech Industry***
3. ***Medical Technology*** *(or Medical device & diagnostic industry)*
4. ***Healthcare Information Technology***

These industries provide the equipment, technology and medicines to the healthcare industry (hospitals, medical and dental clinics and other clinical services) to care for and treat patients.

The increase in world population and prevalence of chronic diseases, combined with escalating research costs, rising global competition and the increasing complexity of distribution systems, are placing unprecedented cost pressures on the biomedical and healthcare industries. Both the biomedical and healthcare industries face the following opportunities and challenges:

1. ***Globalization.*** The U.S. assumed the mantle of leadership of the biomedical industry by being the first to commercialize recombinant DNA research, for example. That achievement was made possible only because it had created an environment and infrastructure in which innovation could flourish. However, if another nation duplicates or improves upon this formula and subsequently makes a pivotal scientific breakthrough in genomics, personalized medicine, embryonic stem cell research, or some other cutting-edge field, it could tip the scales in the other direction.

2. ***Societal, Health and Disease Issues.*** Globalization increases the risk to spread an infectious disease throughout the world. It also facilitates more international collaboration and better disease control. Poverty, lack of clean water and inadequate internet service need to be addressed in under-developed countries. Obesity, chronic diseases, substance abuse and inadequate care of an aging population are some societal and health issues that the biomedical industry needs to work on.
3. ***Science Advancements and Innovations.*** Beginning in the 1980s, the United States surged to the forefront of biomedical innovation. This sudden and remarkable shift was the result of strong policy positions taken by the federal government. The absence of price controls, the clarity of regulatory approvals, a thoughtful intellectual property system and the ability to attract foreign scientific talents to outstanding research universities put the U.S on top. The resulting ecosystem — defined by university-business collaboration, government funding, industry clusters, private equity finance, and entrepreneurship — far surpassed the prevailing European model. The innovative leaps made in biopharmaceutical research, medical devices, and diagnostics gave the U.S. a major advantage that it still holds today.
4. ***Digitization of the Healthcare System.*** With $35 billion in incentives appropriated through the 2009 HITECH Act, the proportion of American hospitals with a system of electronic health records (EHR) has grown eight-fold in recent years, from 9% in 2008 to 76% in 2014. The comparable increase for physicians has been from 17% to 51%. This first step to digitize the American healthcare system captures not only the traditional clinical narrative, but also structured data relating to all aspects of care, including diagnosis, medication, laboratory test results and radiological imaging data. According to a study by the Mayo Clinic, EHRs could be causing physician burnout. More than half of the doctors surveyed (63%) reported that EHRs have failed to improve efficiency, and 41% disagreed or strongly disagreed that EHRs have improved patient care. Now it is up to industry and researchers to improve the interoperability of EHR, simplify the procedure of data input and employ big data and artificial intelligence to analyze the data in EHR so that the care and treatment of patients is made safer and more effective.

1.2. Medical Technology

The medical device industry is one of the biggest industries in healthcare, driven by innovation and new technologies. The last decade has seen an unprecedented growth in innovative and improved technologies, which has led to the development of state-of-the-art medical devices and catalyzed growth and advancement in the healthcare industry (3).

Global sales of medical devices in 2015 were estimated by the Department of Commerce (DOC) to have been $344 billion with the U.S. capturing approximately $148 billion.

The 2014 revenues of the top 20 companies in the world appear in Table 1.1. The device market is dominated by US companies.

Table 1.1. The world's top 20 medical device companies (20).

Company	Country of Origin	2014 Revenue ($ billion)
1. Johnson & Johnson (Devices)	U.S.	$27.5
2. GE Healthcare	U.S.	$18.3
3. Medtronic	U.S.	$17.0
4. Baxter International	U.S.	$16.7
5. Siemens Medical Solutions	Germany	$15.8
6. Philips Medical Systems	Netherlands	$11.2
7. Cardinal Health	U.S.	$11.0
8. Covidien	U.S.	$11.7
9. Abbott Laboratories	U.S.	$10.1
10. Stryker	U.S.	$9.7
11. Danaher	U.S.	$9.4
12. Becton Dickinson	U.S.	$8.5
13. Boston Scientific	U.S.	$7.4
14. Essilor	Switzerland	$6.9
15. Alcon	U.S.	$6.6
16. B. Braun	Germany	$6.6
17. Fresenius AG	Germany	$6.0
18. St. Jude Medical	U.S.	$5.6
19. 3M Healthcare	U.S.	$5.6
20. Olympus	Japan	$4.8
Total		$216.4

There are more than 6,500 medtech companies in the United States. They are primarily small and medium-sized enterprises (SMEs). More than 80% of medical device companies have fewer than 50 employees, and many (notably innovative start-up companies) have little or no sales revenue. These companies employ about 400,000 persons directly and more than two million persons indirectly. The U.S. 2008 medical device market ($88.8 billion) is shown by specialty area in Table 1.2. The cardiovascular device market is the largest segment with a market share of 28% followed by the orthopedics device segment with 19%.

Table 1.2. The U.S. Medical Devices Market by Specialty.

	Market, $ billion	Share (%)
Cardiovascular	$24.9	28%
Orthopedic	$16.9	19%
General Surgery	$10.7	12%
Ophthalmic Equipment	$10.7	12%
Wound Care	$6.2	7%
Hearing & Audiology	$4.4	5%
Other	$15.1	17%

Source: S&P Industry Survey Healthcare: Products & Supplies, 2/4/2010.

1.3. Pharmaceutical and the Biotech Industry

The first drugs were extracts of natural products. However, they are often less potent and broadly active across many body processes. With the progress in chemistry, the biological activities are distilled to reduce adverse effects and increase the therapeutic effects. As human physiology and pathology were better understood, chemicals could be screened, and leads were found that were more and more specific to the biological system. Natural products were replaced by synthetic products. Then scientists studied cells, then cell surface receptors, targets within the cell and, finally, at the gene-protein and protein-protein interactions within cells to generate newer drugs that have fewer side effects, less interactions with other drugs and better efficacy and time duration than older drugs.

The development of these chemicals or drugs leads to the growth of many pharmaceutical companies. These drugs are called "small" because they are much smaller than protein so that they can be taken orally, passed into blood, reached their target in an organ, and produced the desirable therapeutic effect.

The identification of DNA structure by James Watson and Francis Clark in 1953 led to an understanding of the metabolic, physiological and biochemical interactions among proteins in the human body. Within twenty years, many techniques, such as recombinant DNA technology and monoclonal antibody technology, were developed and the biotechnology sector to produce biologics, such as insulin, growth hormone, erythropoietin, was founded and flourished.

The top five pharmaceutical companies (16) and top three biotech companies (15) are listed in Tables 1.3 and 1.4, respectively. The United States is the largest market for pharmaceuticals and accounts for about 35% of the global market, and is the world's leader in biopharmaceutical research and development. The pharmaceutical pipeline has more than 7,000 new medicines in development around the world with approximately 3,500 compounds currently being studied in the United States. More than 854,000 persons are employed by the pharmaceutical industry in the U.S.

Table 1.3. Revenues of the top five pharmaceutical companies (16).

Top Five Pharmaceutical Companies	Country of Origin	2014 Revenue ($ billion)
Johnson & Johnson	U.S.	$70.8
Norvartis	U.S.	$61.1
Roche	Switzerland	$45.5
Pfizer	U.S.	$40.3
Sanofi-Aventis	France	$40.0
Total		$257.7

Many pharmaceutical companies have acquired top biotech companies. However, the biotech company Gilead Sciences, because its Sofosbuvir — the best-selling new drug ever — cures hepatitis C virus, which affects 140 million people worldwide (95% of people with genotype 2 and 56% of those with genotype 3). Gilead's 5-year average sales growth rate (1999–2014) was 15%. The growth rates of Amgen and Biogen were 5% and 10%, respectively.

Table 1.4. Revenues of the top three biotech companies (15).

Top Three Biotech Companies	Country of Origin	2014 Revenue ($ billion)
Gilead Sciences	U.S.	$20.7
Amgen	U.S.	$19.7
Biogen	U.S.	$9.0
Total		$49.4

1.4. Innovations in Medical Technology

"Innovation will be the single most important factor in determining America's success through the 21st century... America's challenge is to unleash its innovation capacity to drive productivity, standard of living and leadership in global markets... For the past 25 years, we have optimized our organizations for efficiency and quality. Over the next quarter century, we must optimize our society for innovation."

This was the opening resolution of an Innovate America report that was presented by Samuel J. Palmisano, Chairman and Chief Executive Officer of IBM Corporation and G. Wayne Clough, President of the Georgia Institute of Technology at the 2004 National Innovation Initiative Summit in Washington DC (11).

On the subject of medical innovations, three papers in the journal *Health Affairs* (5) concluded that:

- Americans believe that opportunities for medical miracles are endless and, thus, are willing to pay for progress.
- When costs and benefits are weighed, technological advances have proven to be worth far more than their costs.
- The ten most important medical innovations developed during the last 25 years, as indicated by a survey of physicians, are:
    1. Magnetic resonance imaging (MRI) and computed tomography (CT) scanning,
    2. Angiotensin converting enzyme (ACE) inhibitors,
    3. Balloon angioplasty with stents,
    4. Statins,
    5. Mammography,
    6. Coronary artery bypass graft (CABG),
    7. Proton pump inhibitors and H2 blockers,
    8. Selective serotonin reuptake inhibitors (SSRIs) and non-SSRI antidepressants,
    9. Cataract extraction and lens implantation, and
    10. Hip and knee replacements.

One goal of American Institute for Medical and Biological Engineering (AIMBE) is to accelerate the growth of the nation's economy and the improvement of healthcare by innovations in medical and biological engineering. Promoting awareness of the contributions made by biomedical engineers and assuring the development of a public policy on a healthy environment for innovation led to the selection of 24 innovations of medical and biological engineering for induction into the AIMBE Hall of Fame. Grouped according to the decades in which the innovations first gained wide usage, they are:

- 1950s and earlier
  - Artificial kidney
  - X-ray
  - Electrocardiogram
  - Cardiac pacemaker
  - Cardiopulmonary bypass
  - Antibiotic production technology
  - Defibrillator
- 1960s
  - Heart valve replacement
  - Intraocular lens
  - Ultrasound
  - Vascular grafts
  - Blood analysis and processing
- 1970s
  - Computer assisted tomography (CAT)
  - Artificial hip and knee replacement
  - Balloon catheter
  - Endoscopy
  - Biological plant/food engineering
- 1980s
  - Magnetic resonance imaging (MRI)
  - Laser surgery
  - Vascular stents
  - Recombinant therapeutics
- 1990s to the present
  - Genomic sequencing and micro-arrays
  - Positron emission tomography (PET)
  - Image-guided surgery

The following three criteria were used by the fellows of AIMBE in their selections from a list of 60 innovations that were nominated for entry to the AIMBE Hall of Fame:

- The innovation must represent a significant engineering achievement

- It must be in general use
- Most importantly, the innovation must save lives and improve the quality of life for a large number of people.

As described earlier in this section, six major medical innovations that were identified by physicians are in the AIMBE)'s Hall of Fame. (The other four are innovative drugs.) To further contrast the significance of these two lists (one by physicians and one by biomedical engineers), we quote from the article by Fuchs and Sox (5) on the physicians' ranked list that
"*The most surprising finding of their study was the extent to which the leading innovations were an outgrowth of physical sciences (physics, engineering, and computer science) rather than disciplines traditionally associated with biomedical sciences.*"

2009 was the 30th anniversary of MD&DI magazine. To celebrate this occasion, the readers of MD&DI chose 30 innovative medical devices that had a significant impact on healthcare during the previous 30 years (1). They are listed below, thanks to the courtesy of MD&DI. The first year in which the device was used in the United States appears first.

1. 1979 Blood and Cell Separator. This device draws whole blood, keeps the desired component and returns the remaining blood components to the donor to eliminate the risk of contamination. It enables donors to give blood more frequently and patients to receive blood from fewer donors.
2. 1980 Implanted Cardiovascular Defibrillator for patients who are at risk of sudden cardiac death due to ventricular defibrillation.

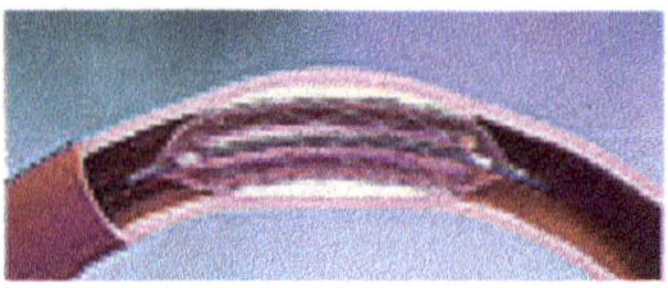

3. 1980 Angioplasty Balloon Catheter This percutaneous coronary intervention has both improved and saved the lives of patients.
4. 1980 Cochlear Implants, which give deaf people the ability to hear.

5. 1980 Intra Articular Arthroscopic Shaver System. This is used in orthopedic procedures to remove bone or cartilage and other soft tissue from a patient's joint.

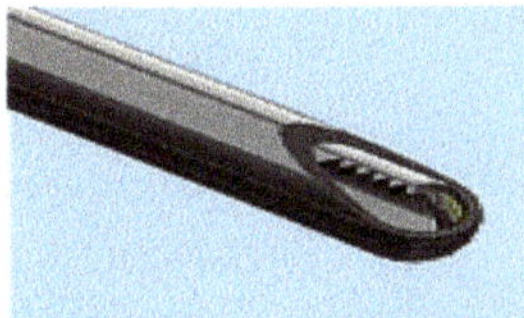

6. 1980 Personal Glucose Meter. This development advances significantly the treatment of diabetes by moving glucose testing from the hospital to the home.
7. 1981 Laryngeal Mask Airway. This ensures that a patient who is under anesthesia has an unobstructed airway.
8. 1981 Pulse Oximeter, a non-invasive way to measure continuously the oxygen saturation level of a patient's blood.
9. 1983 EXCEL and PAB IV Containers. These simple "devices" provide users with infusion fluid that is free of Di (2-ethylhexyl) phthalate. Leaching from the container into the contained fluid may be a critical concern.
10. 1985 Automated External Defibrillator. This has been instrumental in saving lives since its introduction.
11. 1987 Digital Hearing Aid. Its introduction enables manufacturers to enhance features and to provide users with more comfort and higher-quality hearing.
12. 1988 Safety Needle. This contributes greatly to reducting needle-stick, the most frequent cause of blood-borne infections in healthcare setting.

13. 1991 Demineralized Bone Matrix Gel. This is an off-the-shelf product that is used by surgeons for bone healing.
14. 1992 Ventricular Assisted Device. This is a pump to help a weak heart to pump blood through the body and a "bridge implant" to help patients survive until they obtain a new heart.
15. 1992 Smart Infusion System. When it was first introduced, it was the first infusion system to have a dose rate calculator.
16. 1994 Palmaz-Schatz Balloon Expandable Stent. A treatment to solve problems that angioplasty alone could not.
17. 1994 Headless Cannulated Bioabsorbable Interference Screw. Unlike metal screws, its polymer is absorbed by the body and replaced it with bone.

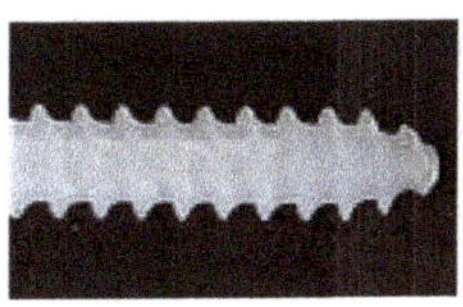

18. 1995 Medical Lasers for surgery. Their most

popular application is LASIK, which is used to correct myopia, hyperopia and astigmatism.

19. 1996 Angio-Seal, which uses bio-absorbable components to seal punctures in the femoral artery after arterial catheterization.

20. 1998 LightCycler PCR (Polymerase Chain Reaction). This is one of the best known devices in point-of-care diagnoses for infectious diseases. The quick disease identification enables immediate treatment and protection for those at risk.

21. 1999 da Vinci Surgical System. Its microchip technology and 3-D optics enable surgeons to perform complex procedures by making tiny incisions to treat in minimally invasive fashion a broad range of pathological conditions.

22. 1999 Cyber-Knife Robotic Radiosurgery System. A miniature linear accelerator that delivers non-invasively concentrated beams of radiation to a targeted tumor. A cumulative dose of radiation kills the tumor cells, while minimizing exposure of the surrounding healthy tissue.

23. 2001 PillCam, a capsule that houses a miniature video camera, lights, a transmitter and batteries. As it is swallowed for passage through the intestine, the device takes photos and sends them to a small recorder that is fastened to the patient's belt.

24. 2002 OraQuick Advance Rapid HIV-1/2 Antibody Test. The test correctly identified 99.6% of people who were infected with HIV-1 and 100% of people who were not infected.

25. 2003 Drug Eluting Stent. The use of medication with the stent has reduced restenosis rates and provided significantly better clinical outcomes for patients. Its introduction ushered in the era of combination products.

26. 2003 LifePort Organ Transporter. This system pumps a cold solution through the organ to reduce tissue damage while it is in transport.

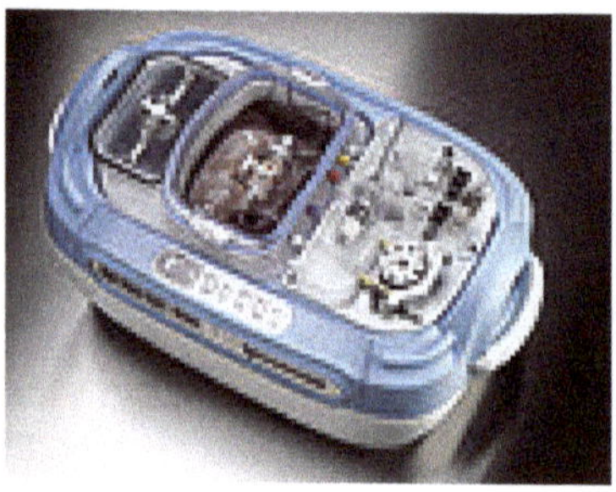

27. 2003 Sidne (Stryker Integrated Device Network) Voice Activation System. This uses voice recognition to

give the surgeon control over endoscopy equipment in the operation room without touching a button.

28. 2005 OxyMask. Its open oxygen design eliminates CO_2 rebreathing and avoids mucosal drying, nose bleeding, facial sores and the claustrophobic feeling that the mask can give. It also enables a patient to communicate more easily and to drink through a straw.

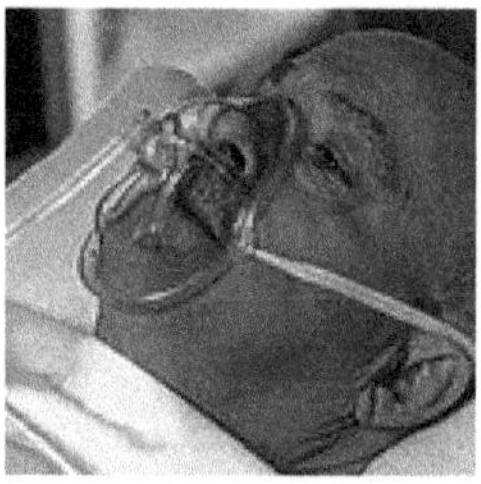

29. 2007 Pinnacle TPN (Total Parenteral Nutrition) Management System. This provides a safe and easy way to check, compound and deliver TPN to patients.
30. 2008 Impella 2.5 Circulatory Support System. This is a catheter that can be inserted into the femoral artery and the heart. The 12F motor and impeller inside the catheter pump blood at a rate of up to 2.5 liters per minute.

The readers of MD&DI also chose the following five "older" technologies:

1. 1800–1970s Hemodialyzers and Dialysis Machines
2. 1960s 1970 Artificial Pacemakers
3. 1972 Computed Tomography Scanner
4. 1973 Vena Cava Filter. This filter has the same basic shape as that used in oil refining to trap sludge and debris, but is used to prevent life-threatening pulmonary embolisms.
5. 1977 Magnetic Resonance Imaging

The following excerpts from the MD&DI article on 30 Years, 30 Devices recount the best aspects of the innovations that have come from the medical device industry:

"Some of the devices nominated by our readers are small enough to travel through a blood vessel. Some are so large they fill an entire room. Some cost thousands of dollars but will stay in the body for 10 years, and some cost pennies and are designed to be thrown away after one use. This industry is characterized by innovators looking for the best way to engineer a solution to a problem. And device designers are noted for their ability to borrow ideas from other industries.

In the last 30 years, we have witnessed the industry's coming of age, with its first blockbuster device and its shift from individual inventors and small start-ups to globally recognized brand names. Here (the above list) is a look back at the devices that have changed the industry and the world."

1.5. Development and Growth of the Pacemaker Industry

Cardiac pacemakers are implanted in approximately five million patients annually. This is equivalent to an implant rate of one pacemaker every six seconds. These patients' hearts have an irregular cardiac rhythm. The use of pacemaker enables the heart in providing the cardiac output that is necessary to supply the oxygen and nutrients and remove carbon dioxide and metabolic wastes from cells. (It is noted that the heart of an adult pumps blood to the peripherals at a rate of five liter per minute or 7,200 liters/day.) To illustrate the inventions, innovations and improvements being made to the pacemaker, we discuss below a number of notable firsts in pacemaker development (2).

- ***The First Artificial Pacemaker.*** This device, which was named by the inventor, Albert Hyman, in 1920s, was a manually cranking device that could deliver electrical pulses at rates of 30, 60 and 120 pulses per minute. However, the use of electrostimulation was not accepted at that time by the medical community and society. Instead, this first pacemaker was roundly dismissed as "gadgetry."
- ***The First Portable Pacemaker.*** This was manufactured by Zoll as a PM-65 that weighed about 45 kg, contained vacuum tubes and was powered by AC. It delivered periodic electric impulses of milliseconds in duration at 50 to 150 volts through a pair of metal electrodes of 3 cm^2 in area, and strapped to the patient's chest directly over the heart. The electrodes irritated the skin and the patients naturally found the repeated electric shocks painful. This pacemaker was used to resolve the heart blockage that developed after cardiac surgery. In a few days, the heart rhythm would recover sufficiently for the patient to be discharged without further use of the PM-65 electrical stimulation.
- ***The First Myocardial Electrode.*** This multi-stranded, braided stainless electrode in a Teflon sleeve was developed in the

laboratory of cardiac surgeon C. Walton Lillehei. One end of the electrode was implanted directly into myocardium and the other end was connected to the PM-65. In this arrangement, 1.5 volt pulse was sufficient to provide effective pacing. After the patient had recovered from cardiac surgery, the electrode could be removed by tugging it out from the body.

- ***The First Battery Driven Transistorized Pacemaker.*** A circuitry that was reported in Popular Electronics was used by Earl E. Bakken, the founder of Medtronic Inc. in 1960 to build this pacemaker. Once, after testing the pacemaker with an animal, Lillehei used it on a boy who was recovering from heart blockage surgery. Later, the pacemaker employed a Hunter-Roth electrode that consumed less power and was less susceptible to mechanical breakage caused by the beating heart.
- ***The First Implantable Pacemaker.*** This device was developed by surgeon Ake Senning and physician-inventor Run Elmqvist of Sweden. It was implanted in a 43-year old engineer, Ame Larsson, in 1958. The patient suffered from Stokes-Adams attacks that required numerous resuscitations daily. The first pacemaker lasted for eight hours. Even though the electrode was replaced five times and 22 pulse generators of 11 different models were used, this patient, who was 86 years of age by then, died of a malignancy that was unrelated to his heart disease or pacemakers.
- ***The First Long-Term Implantable Pacemaker.*** This invention was derived from the work of an electrical engineering professor, Wilson Greatbatch, and two surgeons, William Chardack and Andrew Gage. A U.S. patent entitled "Medical Cardiac Pacemaker" was issued to Greatbatch in 1962. He stated in the patent that the primary objective of the invention was to provide an improved artificial cardiac pacemaker for restoring satisfactory heart rhythm to a heart that is functioning inadequately due to conduction defects in the auricular-ventricular bundle.
- ***Recent Developments.***
  - ***Insertable Cardiac Monitor.*** These small devices, which typically are implanted just beneath the skin of the upper chest, continuously monitor heart rhythms and record them either automatically or when triggered by a hand-held device

to help to diagnose the cause of abnormal heartbeats. This monitor is used to resolve Unexplained Fainting.

- ***Remote Monitoring Systems for Implantable Cardiac Devices.*** These systems link the implantable devices to authorized physicians via the Internet. This enables physicians to perform a complete analysis of all of the data that is stored in a patient's device that would be viewed in an in-office visit.
- **Leadless Pacemaker.** This is delivered to the heart chamber by means of a catheter that is snaked up through the femoral artery. St. Jude Medical's Nanostim or Medtronic's Micra, don't have transvenous leads, are less than one-tenth the size of a traditional pacer, have batteries with an approximate life of 10 years and can be repositioned and retrieved. Both pacemakers have CE Mark approval. St. Jude was expected to file a pre-market approval (PMA) submission for FDA approval in late 2015, whereas Medtronic anticipated having a U.S. launch of Micra by mid-2016. In January 2017, Medtronic secured, under some conditions, Medicare coverage for its leadless pacemaker, Micra Transcatheter Pacing System (TPS).

This brief review of the "firsts" of cardiac pacing demonstrates the exciting history of initiatives, inventions and innovations being made by the collaboration of engineers, scientists, physiologists and physicians. There were criticisms and opposition to overcome. The integration of medicine, technology and clinical trials has developed into a major medical device industry that brought cardiac rhythm therapy out of the labs and into the clinics to benefit millions of patients.

1.6. Three Innovations in Biotech

1. Recombinant DNA. In 1978, Werner Arber, Daniel Nathans and Hamilton Smith were awarded the Nobel Prize in Medicine for creating the rDNA technology. rDNA is made by combining DNA from two or more sources, even of different organisms. First DNA fragments are cut out of their normal position in the chromosome using **restriction enzymes** and then inserted into other chromosomes or DNA molecules using enzymes called **ligases**. The products generated by the rDNA technology are exemplified below.

- ***Recombinant Human Insulin.*** Almost completely replaced insulin obtained from animal sources (e.g., pigs and cattle) for the treatment of insulin-dependent diabetes.
- ***Recombinant Human Growth Hormone*** (HGH, somatotropin). Administered to patients whose pituitary glands generate insufficient quantities to support normal growth and development.
- ***Recombinant Blood Clotting Factor VIII***
- ***Recombinant Hepatitis B Vaccine***
- ***Diagnosis of Infection with HIV***
- ***Golden Rice.*** This variety of rice holds substantial promise for reducing the incidence of vitamin A deficiency in the world's population.

2. *CRISPR Technology.* CRISPR is short for "Clustered Regularly Interspaced Short Palindromic Repeats." This technology was developed just four short years ago by two groups — Jennifer Doudna, a molecular and cell biologist at the University of California, Berkeley, together with Emmanuelle Charpentier, who is now at the Max Planck Institute in Berlin - and Feng Zhang, a biomedical engineer at the Broad Institute of Harvard and MIT. The CRISPR/Cas9 technology was named the 2015 Breakthrough of the Year by Science. Genome editing — a suite of methods for creating changes in DNA more accurately and flexibly than previous approaches — was hailed as the 2011 Method of the Year by Nature Methods.

It is stated in the literature that bacteria use CRISPR as a survival mechanism to fend off infection by viruses. When bacteria were infected, they would surround the virus with viral DNA fragments and then alter the genetic sequence of the virus to make the latter ineffective.

The perfection of CRISPR by these two groups enables scientists to easily and inexpensively find and alter virtually any piece of DNA in any species. According to an article that was written by Time writer Alice Parker (12), the CRISPR is applied to the following technology or product developments:

- ***Mushrooms.*** Scientists have used CRISPR to disable genes that cause mushrooms to turn brown. Because the edits don't involve non-mushroom DNA, the Department of Agriculture (USDA) will not regulate them.

- ***Mosquitoes.*** Since only female mosquitoes bite, some researchers have rewritten the species' genome to produce more males, whereas others have edited the DNA so that the insects are resistant to malaria.
- ***HIV.*** Once HIV infects a cell, it's difficult to reverse the disease process. However, CRISPR makes it possible to splice out HIV from the genomes of cells, rendering them HIV-free again.
- ***Potatoes.*** When cooked at high heat, potatoes can produce cancer-causing acrylamide. Scientists are editing out the gene responsible, making the potatoes healthier when browned.
- ***Cancer.*** In the first-in-human CRISPR treatments, researchers are editing immune cells to make them better cancer fighters and reinfusing them into patients to fight the cancer.
- ***Sickle Cell Disease.*** Sickle cell disease is a monogenic blood disease. Clinical protocol is being developed to correct the sickle cell disease mutation using CRISPR/Cas9 in hematopoietic stem cells.

Two studies (4, 6) published in *Science* and *Nature* in October 2017 broaden CRISPR's reach further. One study extends a strategy for editing DNA, whereas the other breaks new ground by base editing its molecular cousin, RNA (4). Base editing, in contrast, does not cut the double helix, but instead uses enzymes to precisely rearrange some of the atoms in one of the four bases that make up DNA or RNA, converting the base into a different one without altering the bases around it. That ability greatly increases the options for altering genetic material.

Many human diseases are caused by the mutation of a single base. CRISPR has difficulty correcting these so-called point mutations efficiently and cleanly, so base editing could provide a more effective approach. After the *Nature* report of chemist David Liu of Harvard University (6), a group in China used DNA base editing to correct a disease-causing mutation in human embryos cloned from a patient with a genetic blood disorder.

"Gene editing based on nucleases is very good at inactivating genes," says CRISPR researcher Feng Zhang of the Broad Institute in Cambridge, Massachusetts and the senior author of *Science* report.

3. Artificial Cells with Synthetic Proteins. Professor Floyd Romesberg at the Scripps Research Institute in La Jolla, CA said they have engineered the first stable life form that incorporates artificial DNA. The result was published in the Proceeding of the National Academy of Science with Yorke Zhang as the first author and Romesberg as the senior author (21). Other authors are from Grenoble Alpes University, France and Henan Normal University, China.

The artificial proteins can have properties that natural proteins don't have, making them potentially useful for drug manufacturing. By making these proteins in a living organism, the cost to produce medicine can be greatly reduced relative to chemical manufacturing.

In addition, the semi-natural proteins could be incorporated into the organism itself to give it new abilities. Three improvements were made to these proteins:

- The surface molecule used to transport the components of the unnatural DNA into the cell was modified to be more compatible to the cell.
- The synthetic DNA letters themselves were modified to work more efficiently.
- An error-removal mechanism, constructed from a bacterial enzyme called CAS9, a component of the CRISPR/CAS9 gene editing system, was built into the DNA that destroyed replicated DNA that failed to include the synthetic DNA. This ensured that only the full sequence would be replicated.

1.7. The World's Top MedTech Hubs

The top five medtech hubs of the USA (19), the top three hubs in Europe (17) and the world's top three emerging medtech countries (18) as identified by MD&DI are reviewed below.

The first on the U.S. list is Minneapolis/St. Paul where Bakken founded Medtronic Inc. St. Jude Medical has its headquarters in the metro area, and Boston Scientific has a significant presence. 3M Co. is also a major medical device player. As shown in Table 1.5 Minneapolis and St. Paul are unmatched in employment outside of California. The Twin Cities not only have the densest concentration of medical device talent of any metro area in the U.S., but also lead in medical patents and the manufacturing of surgical tools, prosthetics and bandages. Second place Boston has a significant number of medical device startups that

Table 1.5. Medtech employees, VC investment & NIH funding of the top five U.S. medtech hubs (19).

	Medtech Employees	VC Investment $ million	NIH Funding* $ million
Minneapolis/St. Paul	27,718	$246	$505
Boston	18,893	$364	$2,400
San Jose/Silicon Valley	9,372	$983	$1,600
Los Angeles/Orange Counties	29,104	$208	$1,300
San Diego	7,614	$48	$521

* The first two are to their state and the last three to their congressional district.

spun off from research carried out at institutions such as Harvard and MIT. Robert Langer, Ph.D., a notable professor at MIT, has helped to lead a revolution in the field of biomaterials. A slew of medical startups have been spawned by his lab alone. The next three medtech hubs are located in California. A small army of innovative healthcare firms are based in Silicon Valley, which leads the nation in venture capital funding for medical devices and is a hotbed of development of wearable devices. A number of large medical device companies also call the area home. Home to Stanford University, a medical research leader, the region has one of the most educated and talented workforces on the planet. Most of the medical device activity in the fourth hub is based in Orange County and is anchored by such firms as Abbott, Edwards Lifesciences, CareFusion (now merged with Becton Dickinson), and drug-and device-maker Allergan (now part of Actavis). San Diego is emerging as an innovator in the field of wireless medical technology because Qualcomm, the general wireless tech leader, was founded there in 1985 and is helping to foster innovation in the digital health sector.

In Europe, the top three hubs are Grand Est of France, Nurnberg of Germany, and Emilia-Romagna in Italy (17). Grand Est is home to 150 medtech companies. Five new medtech start-ups are established each year in this region of France. The first medical X-ray imaging companies were set up in Nurnberg. Today, the hub has 180 medical technology companies, employing more than 16,000 persons. Some of the technology pioneers in computer tomography, magnetic resonance tomography, interventional diagnostic imaging refractive surgery with lasers, lithotripsy, endoscopy treatment systems, sensors, medical information systems and hi-tech implants can be found in Nurnberg. Emilia-Romagna is one of the most industrial regions in Italy and is home to 392 medtech corporations with 69% of them involved in diagnostics and treatment devices.

China, Australia and Israel are the world's top three emerging countries for Medtech (18). The market for medtech in China has a value of $17.7 billion (2015), but China's medtech exports have a value of $19.3 billion (2013). Those of Australia and Israel are $4.2 billion & $2.23 billion and $1.0 billion & $1.6 billion (2011), respectively. China's medical device market is expected to experience a compound annual growth rate of 7.4% until 2019. More than 19,000 persons were employed in Australia's medtech industry in 2014. A majority of medical device companies in Israel have five or fewer employees.

1.8. Impact of MedTech Innovations on Healthcare

When you review the list of innovations that are described in Section 1.4, you may be impressed by how these innovations save lives and improve the patients' quality of life. On the other hand, you will also read news items that contain these messages:

- The high cost of medical technology: who's to blame? In most industries, technology tends to lower costs. Not healthcare. Why is medical technology out of control and is there any way to curb its cost and spread? Technology and its associated costs account for as much as 50% of medical inflation (7).
- Technology could increase healthcare costs without markedly improving quality, according to experts at Wharton (8).
- Propelled by the increasing use of new drugs, imaging technologies and other wildly expensive innovations, (health) insurance premiums are rising fast (14).

Table 1.6 in the next page was used by Peter Orszag, Director of the Congressional Budget Office in his testimony to Congress to illustrate the factors that contribute to the growth of healthcare spending (10). The contributions from technology-related changes in medical practice range from 38% to 65%. A technical review panel that was formed to advise the Center for Medicare and Medicaid Services (CMS) on future healthcare cost trends used similar information to conclude that about half of the growth of health expenditures is attributable to changes in technology. However, the research that is quoted in Table 1.6 merely suggests that the change in medical practice due to technology advancement contributes about one half of health expenditures. In no way does it support the three news items that were given previously.

Let us examine the meaning of the large percentage (i.e., 38% to 65% in the last row of Table 1.6) by analyzing how an invention is likely to change the medical practice, increase expenditures, save lives and produce higher productivity. The study selected is the implantation of pacemakers in patients done by Moss and Rivers (9) and with the implantation expenses listed by Rinfret et al (13). The productivity of each living patient would be $50,000 per year. Table 1.7 lists the total medical expenditures, total productivity, productivity/expenditures and device cost/total expenditures.

Table 1.6. The estimated contributions of selected factors to long-term growth of healthcare spending per capita, 1940 to 1990 (Source: Congressional Budget Office, 10)

	Smith, Heffler & Freeland (2002)	Cutler (1995)	Newhouse (1992)
Aging of the Population	2%	2%	2%
Changes in Third-Party Payments	10%	13%	10%
Personal Income Growth	11-18%	5%	<23%
Prices in the Healthcare Sector	11-22%	19%	Not Estimated
Administrative Costs	3-10%	13%	NE
Defensive Medicine and Supplier-Induced Demand	0	NE	0%
Technology-Related Changes in Medical Practice	38-62%	49%	65%

These hypothetical calculations certainly indicate that the total health expenditures increase significantly because of the change in medical practice. However, the continuation of the lives of these patients contributes a productivity that is 10 times the expenditures. This benefit is in addition to the fact that there are 39 patients living more than 10 years as shown by the study of Moss and Rivers on pacemaker implantation. Although the price of the pacemaker is much higher than that of the pacemakers of most current uses, the total cost for pacemakers is estimated to constitute only a small fraction (20%) of the total expenditures.

The introduction of expensive medical technology has led to some increases in healthcare cost, although the effectiveness of the technology is still being evaluated. Blaming technology innovations for run-away healthcare expenditures is not justified and does not satisfy our desire to have quality healthcare at an affordable price. As concluded in Chapter 19, it is the responsibility of all constituents, citizens, engineers, physicians, hospitals, industry, professional societies, trade organizations and educational institutions, to work together to utilize innovations and improvements for the good of the nation and the people of the world.

Table 1.7. A comparison of a hypothetical expenditures and productivity for 50 patients with pacemaker implantation and historic heart operation.

	With pacemaker implantation*	With the historic operation**
Total Medical Expenditures	$1,975,029	$400,000
Total Productivity	$19,500,000	$4,000,000
Productivity/Expenditures	9.9	10
Device Cost/Total Expenditures	20%	NA

* 11 patients died during the 10-year study of Moss and River. For simplicity, we assume that these patients were not productive since day one and that the cost of their operation was $18,000 per patient. There were implantation costs and follow-up costs for living patients for a total expenditure of $45,452 per patient. A total of 39 patients lived beyond the study period. The pacemaker was dual-chamber at a unit cost of $7,720. The age of the patients was 68±8 years. For a younger group, the total productivity would be higher.

** For purposes of comparison, Mass and River assume that 80% of 50 patients would receive an open heart surgery to correct the abnormal heart rhythm at a lower cost of $10,000 (in 2008 dollars) per operation. It was assumed that 50% of the patients with this "historic" surgery have a productive life of four years as the surgery might not enable the patient to regain fully his or her normal heart function.

1.9. Three Development Phases of Entrepreneurship

Most medical enterprises are started by persons who have a desire to solve medical problems and a good idea that promises to do so. Unfortunately, desire and ideas alone do not guarantee success. A great deal of hard work and critical thinking is required to get any venture off the ground. In this book, the development of a company is addressed in chapters that were written on the following three development phases:

Phase I. **Assessing your invention**
Phase II. **Launching your venture**
Phase III. **Building your enterprise**

Section II of this book will help working and would-be biomedical engineers to determine how well their ideas will succeed. The three chapters in Section II provide guidelines for potential entrepreneurs to work on the following issues:

- Evaluating what they have and need in order to prepare themselves to become successful entrepreneurs,
- Designing a creditable medical-device invention that will appeal to patients, physicians, hospitals, medical device makers and drug manufacturers, and
- Carrying out meaningful research on the marketing potential of the invention
- Raising funds to conduct a feasibility study with the goal of showing that you can use state-of-the-art technology to build your device. A feasibility study may be interpreted as being synonymous with proof of the concept.

Section III will guide the entrepreneurs in **launching** their ventures to production of pre-production models. The seven chapters in the Section are written to help you with the following matters:

- Describing the financial preparation needed to start a company, as well as building the entrepreneurial team.
- Patenting the invention and maximizing its profit generation potential.
- Advancing the company with governmental support.
- Winning grant money and investment by an angel to carry out the research and development of the medical invention.
- Understanding FDA device regulations and preparing for FDA 510(k) submission of a new medical device,
- Presenting the business plan to attract more capital and executives.

With the tasks of Phase II accomplished, we assume that your company has demonstrated the efficacy of your product and raised sufficient funds to proceed to the manufacturing and sales phases. To provide guidelines for **building up** your enterprise, the six chapters of Session III address the following issues:

- How to manage the company's finances and to price the product.
- How to negotiate employment.

- How to license deals, make mergers and acquisitions and sell the company.
- How to lead the company by effective management of people, time and resources.
- How to get the product manufactured.
- How to establish a market niche for your product.
- How to expand and globalize the company.

References

1. 30 Breakthrough Medical Devices of the Past 30 Years, *MD&DI*, **31:6**, 25 (2009).
2. Aquilina O, A brief history of cardiac pacing, *Images Paediatr. Cardiol.* 8:17–81 (2006)
3. Collins S, A Must-Read Overview of the Medical Device Industry, Nov. 2015, marketrealist.com/2015/11/must-read-overview-medical-device-industry/
4. Cox DBT,..., Zhang F, RNA editing with CRISPR-Cas13, *Science* 25 Oct. 2017.
5. Fuchs VR and HC Sox, Jr., Physicians' views of the relative importance of thirty medical innovations, *Health Affairs*, **20**:30 (2001).
6. Gaudelli NM,..., Liu, DR, Programmable base editing of A•T to G•C in genomic DNA without DNA cleavage, *Nature* (2017).
7. Ham, F.L. Business & Health, Nov. 1989 finarticles.com/p/articles/mi_m0903/is_n11_v7/ai_8540359/
8. Knowledge.wharton.upenn.edu/article.cfm?aricleid=2260, June 10 (2009).
9. Moss AJ and RJ Rivers, *Circulation*, 57: 103 (1978).
10. Orszag PR, www.cbo.gov/ftpdocs/89xx/doc8948/01-31-HealthTestimony.pdf Jan. (2008).
11. Palmisano SJ and GW Clough, National Innovation Initiative, (2004) www.ibm.com/ibm/governmentalprograms/NII%20Final%20Report.pdf
12. Parker A, *The Crispr Pioneers, Times December 12, 116-122 (2016)*
13. Rinfret S, ... Goldman L, Cost-effectiveness of dual-chamber pacing compared with ventricular pacing for sinus node dysfunction. *Circulation,* 111:165 (2005).
14. Weinstein MM, The Nation; Curbing the High Cost of Health, *The New York Times*, July 29 (2001), www.nytimes.com/2001/07/29/weekinreview/the-nation-curbing-the-high-cost-of-health
15. www.biospace.com/News/top-10-companies-leading-biotech/396930
16. www.fiercepharma.com/special-report/top-15-pharma-companies-by-2014-revenue
17. www.mddionline.com/article/seven-most-important-medtech-clusters-europe
18. www.mddionline.com/blog/devicetalk/emerging-medtech-hubs-around-globe-02--05-16
19. www.mddionline.com/top-10-us-cities-medtech-innovation
20. www.mpo-mag.com/issues/2015-07-01/view_features/top-30-global-medical-device-companies#sthash.UMzmSdEo.dpuf
21. Zhang Y, ..., Romesberg FE, A semisynthetic organism engineered for the stexpansion of the genetic alphabet, *PNAS*, 114:1317-1322 (2017).

Chapter 2

Education of Biomedical Engineers

2.1. Advancements in Biomedical Engineering

The past 50 years have seen an explosion of activities in the healthcare field. Medical researchers and entrepreneurs have identified and developed a range of new products and processes that are intended to address an ever expanding range of medical conditions. This growth has been led by the medical device and biotechnology industry and, more recently, the expansion of the molecular diagnostics industry and digital health industry. Although the medical device and biotechnology industry has experienced a number of boom or bust periods during that time, the overall growth in the number of medical products and product revenues has been dramatic.

We anticipate that the coming years will continue to see tremendous growth in the field of biomedical engineering. In particular, those individuals, who are able to marry their clinical and engineering capabilities with an entrepreneurial mindset, have the potential to become this century's Bill Gates or William Hewlett.

During the first three years of the 21st century, important and exciting developments have taken place in education and research in biomedical imaging and bioengineering, including the following:

- The National Institute of Health (NIH) provides Small Business Innovation Research (SBIR) and Small Business Technology Transfer (SBTT) grants to support research and development by small businesses in the United States. 1,330 such grants that totaled $623 million were made in 2015.
- The National Institute of Biomedical Imaging and Bioengineering (NIBIB was established by the NIH, with a 2003 budget of $121 million. By 2015, NIBIB's budget had grown to $328 million. More than 80% of this funded research and training at universities, hospitals, industries and research

institutions throughout the country. Each year, the NIBIB supports more than 800 grants and the work of 5,000 researchers.
- The map of human genetic codes was completed by high-speed sequencing machines.
- The Whitaker Foundation awarded more than $700 million to universities to enhance their educational programs and research conducted by their biomedical engineering faculties and students.

On the industrial front, we also can report these exciting developments and projections:

- The Food and Drug Administration (FDA) has taken steps to streamline its approval process in order to shorten the time required to bring safe and effective medical devices and drugs to market.
- There is a rapid increase in the use of medical devices and prescription drugs.
- The medical and healthcare industry makes up 16% of the 2014 gross domestic product (GDP), and the bio-based industry is projected to account for 30% of the GDP in this decade.

2.2. Establishment of BME Education

The first phase of biomedical engineering research and education occurred in the sixties and early seventies. That was a period of hope and aspiration as this new discipline represents an integration of engineering and biology into one to solve medical problems. Electronics, computers and material engineering were applied to automate clinical laboratories and to create new medical devices. Predictions were made that new medical devices and the use of computer would revolutionize delivery by medical care. Biomedical engineering programs and departments were begun by many universities. The number of such U.S. programs granting undergraduate and graduate degrees was about 40 in 1975.

In the second phase, which occurred during the late seventies and the eighties, it was realized that the revolution in medical care was too ambitious. The biomedical industry was not yet seeking graduates of the biomedical engineering programs. The number of degree-granting programs rose to only about 50.

The 1990s saw the beginning of a third and maturing phase of biomedical engineering education. In this period, the number of degree – granting program, especially at the undergraduate level, has increased drastically, to more than 100 by 2003. This resurgence was caused by the increased understanding of the living system at the cellular, molecular and genetic level, which leads to the development of new technologies and engineering analyses to solve medical problems and to improve diagnosis and therapy. This progress in science, engineering, technology, biology and medicine caused students to become more interested in a career as a biomedical engineer. Many engineers in the biomedical industry refer to themselves as biomedical engineers. In response to this medical progress and change in altitude of the industry, many departments of biomedical engineering are adding genetic, molecular, cellular engineering to their curriculum.

2.3. Supporting Societies and Professional Activities

The Biomedical Engineering Society (BMES) was founded in 1968 to serve as a professional society that represents both biomedical and engineering interests. Its stated purpose is "To encourage the development, dissemination, integration and utilization of knowledge in biomedical engineering" (5). The 8,000 members of BMES include leading researchers at major universities, government agencies and BME corporations worldwide, as well as doctors and industry leaders in pharmaceuticals and prosthetic devices. BMES is the approved leading society for ABET accreditation of BME, bioengineering and technology programs.

In 2017, the BMES Annual Fall Meeting attracted more than 4,200 attendees and had 928 oral and 1948 poster presentations. The meeting theme is ***Engineering Personalized Medicine and Therapies***. Specializations of recent meetings have included: Cardiovascular Engineering, Respiratory Engineering, Biomechanics, Biomaterials, Orthopedic and Rehabilitation Engineering, Neural Engineering, Drug Deliveries, Cellular and Molecular Bioengineering, Stem Cell Engineering, Tissue Engineering, Bioinformatics and Systems Biology, Cancer Technologies, Device Technologies — from Nano to Micro, Biomedical Imaging and Optics, Translational Biomedical Engineering, BME Education and New Frontiers.

In 1991, the American Institute for Medical and Biological Engineering (AIMBE) was established in Washington D.C. The principal activities of AIMBE include acting as an advocate for public policy and disseminating information about medical and biological engineering (4). Its four components are the College of Fellows, the Academic Council, the Council of Societies and the Industry Council. By 2008, the College of Fellows had approximately 1,000 scientists, clinicians and engineers who had distinguished themselves in the practice of medical and biological engineering. The Academic Council is made up of the programs or departments of biomedical engineering of 110 universities. The Council of Societies has 14 society members. (There are more than 50,000 society members who practice medical and biological engineering.). The fourth component, the Industry Council, consists of large and small bioengineering companies.

The year 2009 also marked the 40th anniversary of the Association for the Advancement of Medical Instrumentation (AAMI). In 1969, its founding year, AAMI, under the leadership of Dwight Harken, organized a National Conference on Medical Device Regulation in Bethesda, MD to address device safety concerns of the medical community and the government. As John Abele, AAMI founder and Board Member, noted, "This conference was a huge event. It was the first time that the presidents of every American medical association were present in one room." The recommendation of the task force on device regulation that emerged from this conference formed the framework of the "Medical Device Amendments" legislation that was passed by Congress in 1976.

The Advanced Medical Technology Association (AdvaMed) was formed by companies that produce medical devices, diagnostic products and health information systems. It is a trade association that was formed to promote policies that foster the highest ethical standards, rapid production approvals, appropriate reimbursement and access to international markets. AdvaMed's 2009 priorities are to foster continued innovation and to ensure delivery of the best possible healthcare for all Americans.

The Medical Device Manufacturers Association (MDMA) is a national trade association that provides educational and advocacy assistance to innovative and entrepreneurial medical technology companies. It has been a voice for small companies and plays a proactive role in helping to shape government policies that have an impact on medical device innovators.

In addition to the professional magazines that these societies publish, there are a number of prominent magazines that serve the medical device industry. The premium magazines for the biomedical engineering community and medical device industry include BMES' Annals of Biomedical Engineering, AAMI's Biomedical Instrumentation & Technology, IEEE's Engineering in Medicine and Biology Magazine, MX: Business Strategies for Medical Technology Executives, MD&DI (Medical Device and Diagnostic Industry) Magazine, Medical Device Link, Pharma Med. Device, Nature and Biotech.

2.4. Career Interests of Biomedical Engineering Students

The top career choice of 401 biomedical engineering (BME) undergraduates 10 to 20 years after their university graduation is entrepreneurship. It is time to not only recognize this, but also to modify our curriculum so that it can prepare students to meet their career interests and expand the biomedical engineering industry.

Advances in electronics, computers and materials have contributed greatly to the development of the medical device and diagnostic industry. The growth in this industry leads not only to a better understanding of biology and medicine, but also to better healthcare. This industry differs from a consumer-oriented industry.

First, most products of the biomedical industry serve a smaller subset of people, as exemplified by the use of the glucose sensor for those who have diabetes. However, as we acquire a better understanding of health and disease, this industry will be ready to produce many new devices, diagnostic methods and treatments for the prevention and alleviation of a wide array of diseases. Thus, the biomedical industry is fertile ground for BME graduates who wish to assume a leadership role in the development and growth of BME ventures.

Many of the brightest university students have chosen to study BME. This certainly is an exciting field in which to conduct leading research. They will be well rewarded as their work will be able to save lives and improve the health of patients. The author believes that, if the new discipline of BME is to mature into a full-blown engineering discipline, it must have more entrepreneurs and jobs for our graduates. For this reason, I undertook a survey of the career choices of 401 BME undergraduate students at the University of California San Diego, Johns Hopkins University and the University of Southern California. I wanted to learn whether these undergraduates were interested in becoming BME entrepreneurs immediately after graduation or 10 to 20 years later when they had gained more professional experience and education.

As an introduction to the career survey, a brief lecture entitled "Challenges and Opportunities in BME Entrepreneurship" was given to students who were taking BENG 1 "Introduction to Bioengineering" (UCSD, Instructors: Shu Chien and Peter Chen), BME 202 "BME in the Real World" (JHU, Instructors: Sasha Popel and Art Shoukas) and BME 101 "Introduction to BME" (USC, Instructors: Jean-Michael Maarek and Jesse Yen). Freshmen were 63% of the respondents. Sophomores, juniors, seniors and others made up 9%, 12%, 11% and 5% of the sample respectively.

In the lecture, I highlighted the challenges and opportunities in BME entrepreneurship by quoting from an article that appeared in the January, 2006 issue of MD&DI magazine. It says that ***the implantable medical device industry has grown exponentially over the last thirty years***. When a large number of patients are using a medical device, the industry has large revenues and employs many engineers. On the other hand, this quote also implies that a much longer time is required for patients to accept medical devices than is required for consumers to accept a new type of television. Should a medical device cause a few patients to experience a serious incident, it may lead to the end of that device company.

Several medical devices were discussed to further highlight the challenges and opportunities. For example, total knee replacement has enabled patients to regain motion without pain. However, about 10% of implanted total knee replacements may become defective in 10 years. The annual Medicare cost of defective knee replacements amounted to $208 million for the year 2005. One of every 200 patients dies within one day following knee replacement. The message to the students is that there is still a great need for them to improve the engineering of total knee replacement and its implant procedure.

In 2005, one million cardiac pacemakers were implanted in patients worldwide. Medtronic and Guidant (now part of Boston Scientific) are major companies that produce many pacemakers. They had 2005 revenues of $10 billion and $3.8 billion respectively. Medtronic was founded as a two-person company in 1949. The precursor of Guidant began as a three-person company in 1972. The message is that many large medical device companies had humble beginnings and required many years to succeed.

At the end of the lecture, 3" × 5" index cards were distributed to students during the discussion period on which to recording their names, class years, majors and minors. The following six career choices for the students to prioritize were listed on PowerPoint:

1. Industry
2. Government
3. Graduate or Professional School
4. University or Research Laboratory
5. Entrepreneurship
6. MD or LLD.

I advised the students that, if they choose Category 5 Entrepreneurship, it would mean that they would work on an entrepreneur team to build, organize and manage a business venture. The selection of Category 6 MD or LLD would mean that they would be clinicians who work with patients or attorneys who work with clients. If they obtain an MD and wish to teach at a university or conduct research in a research institute, they should choose Category 4.

I also reminded the students that employees in industry can form an entrepreneurial team to work on projects that are an extension of the company's work. Further, industrial jobs are more stable. Many small ventures will fail during their first three to five years. On the other hand, the entrepreneurs own the companies, whereas employees of a large company may own only a small share.

The first question asked the students to identify their first and second choices of six categories of career choices to pursue upon graduation from university. By ranking the first and second choices, I obtained the distribution of their career choices, which are depicted by the open bars in Fig. 2.1 of the next page. One can see that 45% of these students indicated that they would choose to attend graduate or professional school. Many wanted to go on to teach at a university or to practice as clinicians or attorneys. However, 20% chose being employed in industry right after graduation. Only 6.5% thought that they might work as entrepreneurs.

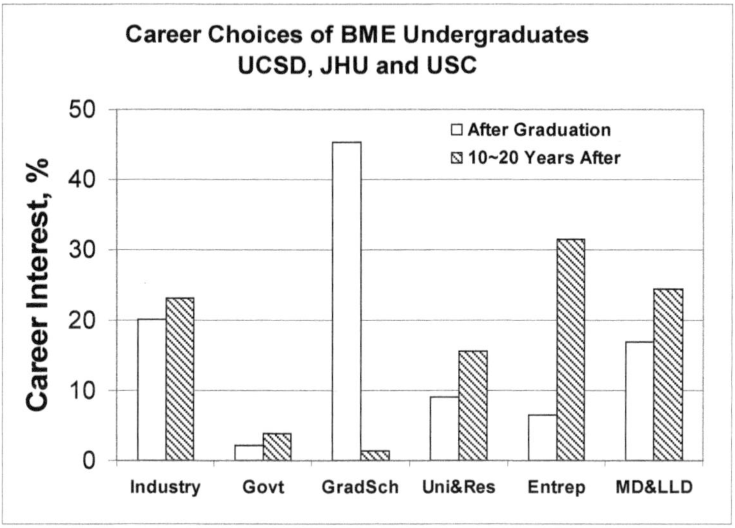

Fig. 2.1. Career choices of BME undergraduates of the University of California San Diego, Johns Hopkins University and the University of Southern California (2).

Similar two-choice answers were obtained for their career interests 10 to 20 years after graduation. They appear as shaded bars in Fig. 2.1. By that time, few would be thinking of teaching or conducting research at a university. Instead, the top choice is entrepreneurship at 32%. The second is the MD or LLD category (25%) and the third is Industry (23%). These distributions are quite similar for each of the three universities, as the standard errors of the distributions are less than 2%. The responses seem to indicate that the students had given considerable thought to their future career well before my lecture and understood the challenges and opportunities in BME entrepreneurship. Based on my informal discussions with the students, I believe that those who chose an entrepreneurial career really want to use their BME knowledge into inventions and then commercialization for the benefit of people.

2.5. Knowledge Preparation for Engineers, Scientists and Physicians

In the next thirty to fifty years, the world will be running out of clean air, clean water, oil and many material resources. It is the mission of the people, especially the scientists, engineers, physicians and many professionals, to assure that the world is still prosperous and that the people live healthy lives. In the short term, it is the responsibility of the scientists to gain a better understanding of the physical world and living systems, of engineers to develop technologies that are essential to the health of the world and living systems, and of physicians to make more accurate diagnoses and to treat or cure diseases more safely and effectively.

Being a scientist, engineer, physician or similar professional will give you a chance to improve the world. In this and the next section, we examine the education process from which you can become a professional who is highly sought by your future employer.

In high school, you should be taking courses in science (including biology), technology, engineering, mathematics and biology. Read several introductory books on medicine and a few biographies of scientist, engineers and physicians. It is desirable to participate in extracurricular activities to improve your communication skills and to enhance your citizenship. The earlier you know the meaning and significance of patent, improvement, innovation, business operation, marketing, etc.; the better prepared you will be to become an entrepreneur.

If you are admitted to an engineering school, you probably will need 120 to 130 semester hours of course work to earn a B.S. degree in engineering. In many schools, the freshmen and sophomores take similar science and engineering courses. After completing those courses, you can then select the engineering specialty in which you will major. You also can select a minor and take a few elective courses in that discipline. Use the elective option to take several humanity courses and participate in class or work projects to strengthen your interpersonal and communication skills.

When your engineering program is accredited by ABET or in a reputable research university, your graduation from that program means that you have satisfied the educational requirements for the profession and are technically qualified to do the engineer's job in the discipline that you majored in.

In a 2014 report by Jamie Hartford for *Medical Device Business,* the highest paying jobs in medical technology are identified as:

- Computer hardware engineer (with an average annual salary of \$101,720)
- Biomedical engineer (\$98,260)
- Chemical engineer (\$96,630)

- Electronics engineer ($96,040)
- Electrical engineer ($92,480)

This report describes a biomedical engineer of medtech as one who "applies knowledge of engineering, biology, and biomechanical principles to the design, development, and evaluation of biological and health systems and products."

It is noted that the salaries are not statistically different. However, the data indicate that the Medtech field pays well not only to biomedical engineers, but also to engineers of the traditional disciplines.

When you enroll to a biomedical engineering program, you will take engineering courses that are oriented toward solving biomedical problems. These courses may span the spectrum of mechanics, electronics, chemical engineering, computer science and material engineering. You will have a two-semester course on physiology and related topics. The teachers will be mostly faculty members of the biomedical engineering program who are using engineering to gain a better understanding of the living system or technologies to better diagnose or treat the patient. Thus, you will have exposure to the medical problems that you may someday build your entrepreneurship on.

The weakness of a bachelor biomedical engineering program is that the graduate will lack an in-depth knowledge of traditional engineering disciplines and adequate problem solving skills. One remedy is to continue one's education to an M.S. or Ph.D. program. In most M.S. programs, the student will invest a significant amount of time to conduct research and prepare and defend a thesis. Conducting the research is equivalent to solving a problem, writing a thesis to describing the problem and solution and defending the thesis to advocate your solution before your peers. In essence the knowledge and experience that you gain from conducting research and thesis writing will elevate your problem-solving ability.

2.6. Character Preparation for Engineers, Scientist and Physicians

To meet these responsibilities, scientists, engineers and physicians who are interceded in building a biomedical engineering enterprise must learn and possess the following five characters or skills:

1. ***Wise, Competent, and Good Communication.*** The technical aspect of wisdom and competency are achieved through the

education that was examined in the previous section. However, the human relation side and communication skills are learned in and outside of the classroom. Good communication is important for you in recruiting scientists, engineers and physicians to work in your company, to make your team productive and to obtain investment capital for your company. Good communication involves listening and genuinely respecting others' opinions. Whenever you work with a group, you should expect to encounter disagreement. You should embrace dissent though your wisdom.

At present, China is the most populous country in the world. Its GDP by purchasing power parity (PPP) is ranked by the International Monetary Fund (IMF) as the first in the world. The annual rate of growth of China's GDP is 6% to 7%. About half of the expected world GDP growth from 2017 to 2019 will come from the US and China. China's trade with the world will be enhanced by their One Belt and One Road Initiatives. Being fluent in Chinese should be pursued as part of the effort of becoming a good communicator.

2. ***Honesty, Integrity, and Ethical Behavior.*** Only if you possess these character traits can you be trusted by the employees and stockholders of your company, the customers or users of your products, the scientific community, and the government and the public.
3. ***Caring for Your Team Members, Coworkers, Your Company, Your Customers and Patients.*** This is about inspiring people, serving people, caring for people, and caring about people. You must show that you care not only by your daily actions, but also by the way that you price your products, set your own salary and stock options and use your wealth.
4. ***Passion, Energy, Courage and Decisiveness.*** It is easier to be a follower and to let someone else solve a problem or make a decision. It's far more difficult to stand up, speak up and be accountable. Be passionate, energetic, courageous and decisive in what you are doing. Warren Buffett terms energy as someone who takes initiatives. Under-promise and over-prepare. One of the most gratifying rewards that you can earn is your team's recognition that you truly are part of the Team.

5. ***Good People and Business Management.*** Teamwork isn't about going along. Things are not accomplished by orders. Teamwork involves hearing all views, setting up team rules, admitting mistakes and sharing risks and rewards appropriately. Good management can solve social problems skillfully.

By knowing which school you graduated from, the person who is conducting the job interview will know if you have the knowledge to do the job. If you are shown to have the five character traits mentioned previously, the company will know that you can get the job done. This will put you into an excellent premium position to find a job that you like and is well paid. Later, these characters traits will prove to be essential in building your entrepreneurship.

2.7. Robert Langer and 31 Start-ups

Tissue and organ failure that are produced as a result of injury or other type of damage is a major health problem, which, accounts for about half of the total annual health care expenditures in the US. Treatment options in the 1990s included transplantation (human or xenotransplantation), surgical repair, artificial prostheses, mechanical devices and, in a few cases, drug therapy. Ultimately, however, major damage to a tissue or organ can never be repaired or long-term recovery effected in a truly satisfactory way by these methods.

Since the 1990s, tissue engineering has been emerged as a significant possible alternative or complementary solution, in which tissue and organ failure is addressed by implanting natural, synthetic or semisynthetic tissue and organ mimics that are fully functional from the start, or that develop the required functionality. The initial efforts have focused on skin equivalents for treating burns, but an increasing number of tissue types are now being engineered, as well as biomaterials and scaffolds that are used as delivery systems. A variety of approaches are used to coax differentiated or undifferentiated cells, such as stem cells, into the desired cell type. Notable results include tissue-engineered bone, blood vessels, liver, muscle and even nerve conduits. Due to the medical and market potential, there is significant academic and corporate interest in this technology.

In the 1980s, R&D in tissue engineering and biomaterials took off, as evidenced by the establishment of tissue engineering programs at major universities around the world and the report by Nature Biotechnology that some 24 companies in the U.S. are developing tissue engineering reports.

The rapid growth in research and entrepreneurship in tissue engineering can be illustrated by the research accomplishments of, and engineering companies founded by, Professor Robert Langer of MIT, an academic researcher who pioneered this new field of tissue engineering.

Langer runs the largest tissue engineering laboratory in the world with an annual budget of more than $10 million at MIT in Cambridge, Massachusetts. From his lab, he has launched 31 companies with students and collaborators and secured 1,100 patents, which have been licensed to another 300 pharmaceutical, biotech and medical device companies. He started his first company, he wrote in a 2013 *Nature Biotechnology* article (1), "because I realized it was an effective path for transforming science into life-saving and life-improving inventions." He owns shares in the 31 companies that are valued at $120 million. He has written more than 1,300 scientific papers and is the world's most-cited engineer. He is also one of only 11 Institute Professors at MIT, the university's highest honor.

In recognition of his work, Langer has won virtually every major engineering and technology prize in the world. President Obama awarded him the National Medal of Technology and Innovation, the country's top honor for scientists in 2013. Two years later, Langer added the prestigious $1.5 million Queen Elizabeth Prize for Engineering to the list. The Queen Elizabeth prize committee estimated that he has improved at least two billion lives by his discoveries.

Langer's entrepreneurial success is characterized by Ann M. Thayer of the American Chemical Society as "One chemical engineer, 30 years, and 31 start-ups" (1). Among the companies that are connected to Langer, nine have become publicly traded corporations, raising more than $300 million combined by IPOs — and more by follow-on offerings. According to the National Venture Capital Association

(NVCA), about 200 biotech and medical device firms receive initial venture capital funding in any given year. That Langer has been able to start up at least one company, if not two or three, in any year is a tribute both to the productivity of his lab and the value that investors assign to his work. The venture capital firm Polaris Partners has invested in at least 20 of Langer's companies and spent more than $220 million in doing so. Noteworthy also is that just a handful of Langer's companies have gone bankrupt or shut down, a record that far exceeds expectations for venture-capital-backed firms. Among biopharma and medical device start-ups, "more companies fail than succeed" and often within a few years, according to the NVCA.

The scope of the technology of these 31 startups can be illustrated by the following four companies:

- ***Neomorphics.*** This was founded in 1999 with a $6 million venture funding. Neomorphics own the patents of the platform technology for tissue scaffold culture. It was acquired by Advanced Tissue Sciences for $21 million in 1992.
- ***Microchips Biotech.*** This was founded in 1999 on a silicone-based drug delivery technology and raised $58 million in four funding rounds. One of its prototypes is a star-shaped object that is housed in a dissolvable capsule. When a patient swallows the capsule, it dissolves and the prototype expands in the stomach. Its odd shape can't pass through the digestive track. So, the device remains in the stomach and releases medicine slowly over time. This type of drug delivery is particularly useful to patients in developing countries with limited access to medicine. That's why Bill Gates once asked Langer to invent an implant that would provide a steady release of a birth control agent and permit her to switch the doses on or off. MicroChips plans to debut such a device in 2018.
- ***Eisai Inc.*** This Japanese pharmaceutical company bought the licensing right to a wafer that contains carmustine, a nitrosourea alkylating agent, distributed in a biodegradable copolymer matrix. Implanting the wafer into the brain following surgical removal of a malignant glioma permits direct delivery of chemotherapy to the tumor site.
- ***Moderna Therapeutics.*** This was founded in 2011 and led by Stephen Bancel. Moderna is a company that develops

messenger RNA therapeutics. The company's technology uses mRNA made of nucleotide analogs to trigger the body's natural processes to produce proteins inside the human cell. This approach has the potential to produce therapeutic proteins *in vivo* to treat a wide range of diseases, including those that cannot be treated effectively by existing methods, to develop and manufacture drugs. Since 2012, Moderna has raised $650 million in venture capital. The company operated "below the radar" for a few years, but is now valued at more than $3 billion. Moderna's unprecedented rate of growth can be attributed in part to how the company leverages its novel platform technology to the pharmaceutical industry.

Langer spoke with *International Business Times* in 2015 on entrepreneurship and the pricing of medical products. The questions and answers as reported by Amy Nordrum appear below (3):

- How do you know when it's time to launch a new company for one of your projects? Before I start a company, I want it to be reasonably advanced. I want to feel that we have a shot at it. We've developed it to the stage of a good solid paper or papers. We probably have animal data that proves the concept. We have a patent. Also if it's going to be a company, it usually is a platform technology that you can use for many. Do you have a platform technology, a seminal paper and a blocking patent? If not you may be in trouble.
- What must an MIT student do to earn a spot in your lab? We have about 5,000 applicants a year. So, it's difficult. I look for superstars. How do I know? I rely heavily on what people whom I know tell me. If people at Harvard or Stanford say that this is the best person, I listen to that. Five to 10 are accepted every year.
- As a biotech veteran, does the growing pushback against the high price of medicines concern you? I understand the issue of price, but when price controls come up, it has a chilling effect on what we're trying to do and what we're trying to get out into the world. It's difficult to get investors. They could invest in the internet or Uber. Why should they invest in us? I'd rather see investments go into medicine. I think that, in the long term, it

will probably be fine, but in the short term, it's not good because at all of the companies, as well as our labs — are always trying to raise money. In some places, you can raise money on sales. Here, you're raising money on hopes that often can be fulfilled, but unless you get the money, you have no chance to do it. It's not as if these things are easy to do.

- What do you think are some of the most common mistakes that other professors and biotech entrepreneurs make when they try to do this? I think one is starting too early. I do feel that in the medical area, it's easy to fail. There's so much money required and it takes so much time. The second reason is that you may not be working with the right people. There will be some investors who are just better than others. The third — and this is the most difficult — is finding the right CEO. When people ask me why some companies have performed better than others, the number one reason is the CEO. I actually don't begrudge CEOs for getting so much money, even millions of dollars. They're worth it. For me, the CEO has been the single biggest differentiator.
- What do you look for in a CEO? The CEOs are smart people and able to pick things up in different ways. I guess I want to be sure that I like them. I want to make sure that they are driven. I would like them to be charismatic. To me, there's almost the optimization issue of selling. You don't want to over promise because that's bad and you don't want to under promise either. You want others to share your vision that your company could change the world and, for that to happen, you must believe it.

References

1. Langer R, A personal account of translating discoveries in an academic lab, *Nature Biotechnology*, 2013, 31:487–489.
2. Lee, J. S., *Bulletin Biomed. Eng. Soc.*, 30: 4, p.17 (2006).
3. Nordrum A, Robert Langer, top MIT biomedical engineer and father of 30 companies, on how to launch a successful biotech company, *International Business Times*, 10/14/2015.
4. www.aimbe.org
5. www.bmes.org

Chapter 3

Evaluating Your Entrepreneurship

3.1. An Entrepreneurial Team

Webster's dictionary defines an entrepreneur as someone who organizes, operates and takes risks in a business venture in the expectation of gaining a profit. Do not interpret this definition to be the lone-wolf model in which the person goes out and does all that is required to start a business in relative isolation. For high-tech businesses, very little is accomplished without a team effort and no one can master all of the skills required to establish and operate a biomedical engineering venture. The person who is in charge of the team must have leadership qualities.

Leadership differs from managership. The former attempts to instill a vision, trust and self-confidence in the team, whereas the latter emphasizes the management and control of the team.

If you want to become a successful entrepreneur, you must possess another critical attribute - perseverance. The ability to stick with a task for as long as it takes to complete it is essential if you hope to raise venture capital, move your product to the manufacturing and marketing phase and make the first few sales.

You will seldom have an opportunity to present your business plan to a venture capitalist as the result of a single phone call or an introduction from an acquaintance. In fact, the venture capitalist may intentionally put you off for a while in order to see how well you persevere in the face of delays or an initially lukewarm reception. Perseverance is a trait that the venture capitalist will want to see you demonstrate before he or she grants you an appointment to make a 10 to 20 minute presentation of your business plan.

Similarly, making the first sale of your product, which is an equally challenging task, can be done only with perseverance. The perseverance of an entire entrepreneurial team that is committed to developing and selling a high-quality product is essential if you want your product to penetrate the market significantly, particularly in the face of competition.

What additional trait does it take to become an entrepreneur in the medical device industry? One aspect that's essential is commitment and desire, — a passion, if you will, to explore unknowns to improve the quality of life of patients. This is not unlike the passion and curiosity that drives researchers to engage in scientific endeavors. Since it is easy to become discouraged by set-backs and the belief that most ventures will fail, the ability to pick yourself up from a setback, dust yourself off, and begin all over again with courage is vitally important for an entrepreneur's eventual success.

In essence, the entrepreneurial team needs a leader, who has:

- Leadership
  - Builds up the team with qualified and motivated individuals,
  - Inspires and orients all with the vision and goals of the company,
  - Generates the team's trust and self-confidence,
  - Manages the team for timely task execution and device development,
  - Gears up the sales and marketing of the products,
- Perseverance
  - Remains persistent despite an expectation that 80% of ideas or proposals will be rejected,
  - Raises funds to establish and operate the company,
  - Gets the product developed to its utmost level in performance, safety and effectiveness,
- Courage and Passion
  - Takes calculated risks as the company develops and markets its product,
  - Makes changes in the company's direction when warranted,
  - Does things ethically and professionally with the goal of benefiting people.

These traits of a team leader are summarized in Fig. 3.1.

One popular myth about entrepreneurs is that they are basically risk-takers. This is not true. Entrepreneurs are risk *managers*. They are very good at figuring out the minimum essential risks that they will have to take to get the company off the ground, and cleverly devise ways to share these risks with other persons.

For example, entrepreneurs may start a company with their own money, but, as soon as they can, will borrow money from banks and investors so that they don't have to use more of their own money than is necessary. They will develop contracts and enter into business arrangements that require customers, vendors, and other people to bear some of the risks.

Fig. 3.1. Three major attributes that the leader of a biomedical engineering entrepreneur team must have.

Entrepreneurs who perceive themselves to be risk-takers are actually gamblers and, in the long term, the house always wins.

Something else that entrepreneurs do well is to determine what might appear to be a risk, but really isn't. They will dig around to discover that the venture is really quite different than what one might perceive from the outside. By cutting right to the heart of the venture, they can determine the minimum essential work to do and the minimum essential risks to take in order for venture to achieve success.

Entrepreneurship is an art that can be learned. Think about it in this way: In medical school, professors teach their students the best scientific practices for use in healing sick patients, as well as more subjective things like an appropriate bedside manner. However, there are limits to what any professor can teach. To develop into a capable physician and provide the best possible patient care, a medical student must develop the art of getting to the heart of an issue.

The same is true for the entrepreneur. An entrepreneur can be taught the best business practices and develop more subjective attributes like strong interpersonal skills, but he or she must also develop the ability to get to the heart of an issue.

Since entrepreneurship is a team effort, you have two options in an entrepreneur team — to lead or to follow. If you are inclined to lead, you can create a venture and do whatever you choose. If you are inclined to follow, you can join an entrepreneur team and support the growth of the new venture.

For example, a technically oriented engineer who has an inclination to follow will function well in a technology group that follows the company's mission by producing a product that is significantly superior to that of competitors.

There are also opportunities for biomedical engineers who wish to work in a new venture's business group. The roles will include those of sale associates, accountants, and lawyers, all of whom support the entrepreneurial leader in his management of the company's operations.

Regardless of the role that you choose, the goal of the entrepreneur team is to formulate a business plan, to raise the capital, to develop the technology and products and, finally, to manufacture and market the products of the company. The team is like a salesman who has the vision, perseverance, passion, courage, arrogance, and belief that they can run the company and sell the products.

3.2. John Abele, Pete Nicholas, Medi-Tech, and Boston Scientific

A partnership arrangement, which operates by consensus, seldom works for a developing company. However, the complexity of running a high-tech business may require a team of technology and business experts to work together. A good example is the partnership of John Abele and Peter Nicholas, who formed Boston Scientific to acquire Medi-Tech, a company that was making steerable catheters for percutaneous transluminal angioplasty and cytology brush for lung sample collection, in 1979. As described in *The Ship in the Balloon*, a book by Jeffrey Rodengen (3), each of these two partners brought something unique to the company. *Nicholas, with extensive managerial and finance experience, would be responsible for the business decisions. Abele, with his enduring interest in the medical technology that was slowly gaining ground, would enhance the cause of minimally invasive therapy and guard the soul of the enterprise.*

Here are three excerpts from Nicholas on this partnership:

"We are simply going to run our business on the basis of what we agreed to agree about and what we agreed to disagree about, and that's it."

"We are going to develop a partnership. We are going to create a 50/50 enterprise. We made a commitment to make this little enterprise the sole focus of our careers."

"I was thinking about ownership from a control point of view only because I didn't know him (Abele). As I got to know him, it became clear to me that company control was not the issue. We were so close in philosophy that he would almost always agree with what I wanted to do."

In April, 1979, Cook Inc. expressed an interest in acquiring Medi-Tech for $400,000 in cash with the condition that Abele would stay on as President for two years. To make a bid for the company, Abele wrote a letter to Cooper Laboratories, the owner of Medi-Tech in which there was the following statement:

"*Until now, the only other parties who are interested in Medi-Tech were interested in it only with me, and only if it were operated as a development group. I am not interested in that arrangement……*

The only reason I am in a position to make any offer is that a gentleman by the name Pete Nicholas has agreed to work with me on a partnership basis. He brings not only financial backing and expertise, but the critically needed administrative, strategic and people skills that complement my technical and market strengths. He is motivated by the same desire for equity and independence as I am. He is not interested in working for Cooper. "

At that time, Medi-Tech's annual revenue was $2 million. According to Rodengen's book, a sale price of $800,000 was reached by negotiations that were long and complex. Nicholas persuaded some bankers to put up $500,000 and the partners indebted themselves to the maximum in order to scrape up an additional $300,000. They turned away several investors and investment firms in order to keep Boston Scientific a private company. Immediately after the takeover, they reorganized the company for long term success by aligning the company's efforts with these company's strategic intents: increasing R&D spending, expanding sales coverage, acquiring new business units and creating a company in which culture, mission and values reigned supreme. In 1992, Boston Scientific went public and Abele and Nicholas sold 25% of the company for $450 million. As a result of their effort and many more acquisitions, Boston Scientific's 2016 revenue reached $8.4 billion.

3.3. Nature of Entrepreneurship

Is an entrepreneurial venture an independent venture? In some cases, the answer is yes. In others, it is no.

Some entrepreneurial ventures are launched within large existing corporations. A large corporation can create a new, partially-independent, ventures division to investigate new market opportunities, and to pursue new businesses development. The corporation may provide internal funding for its employees to explore the potential of a venture. Encouraging results may lead to more company support to conduct the research and development of this partially-independent venture.

Another type of entrepreneurial venture is known as an acquisition. In this situation, the entrepreneur buys an existing small business, often a young start-up. The entrepreneur then decides how to add substantial value to the acquisition in excess of what the previous owner had dreamed of, or incorporates the acquisition into a product that he or she is already developing in order to significantly enhance that product's function.

Regardless of the form that a new venture takes, independent, part of a large corporation, or an acquisition, entrepreneurship requires innovation. Innovation has not only a technical side, but also a business side. As an entrepreneur, you will be called on to manufacture new products, open up new markets, acquire new customers, and develop new ways of thinking about operating a business. The venture usually starts small and grows rapidly to become large. So, high-tech entrepreneurship is about new business, rather than small business.

As a biomedical engineer, are you in the right profession to become an entrepreneur? There are many traditional industries that are involved in mechanical, civil, chemical, and electrical engineering. New computer engineering and information technology industries have grown rapidly over the last 30 years. Now, the time is ripe for industries that are based on biomedical engineering to be launched and developed. So, yes, you are in the right profession.

Will it be easy? Not necessarily. Just like the stock market, it can be very difficult to pick marketable innovations that will become winners — the kinds of exceptional and useful products on which a biomedical engineer might hope to build an entrepreneurial career.

Even so, the field of biomedical engineering offers its practitioners the opportunity to engage in a high-quality career with many professional options, rather than a low-quality career with few options at all. In the opinion of both the author and Wendell E. Dunn III, an entrepreneurial education expert who previously served on the faculty of the University of Virginia's Darden School of Graduate Business and is now Foundation Chair in Entrepreneurship at the Business School of the University of Auckland, the nice thing about medical devices and biotechnology is that they will have a significant impact on healthcare for our society and the world — during the remainder of our lives, in the same way that information technology has had on society over the last three decades and will continue to do long into the future.

We're coming now to the era of medical devices and biotechnology. With the tremendous progress that has been made in basic biological research, medicine and engineering, we will see extraordinary development in the field of medical device and biotechnology in the next 50 years. This will be particularly evident in biomedical engineering innovations, clinical implications, spin-outs, commercial opportunities, and everything else that accompanies biotechnology and biomedical engineering.

3.4. Wanting to Be an Entrepreneur

Is entrepreneurship equivalent to small business or self-employment? The answer is both yes and no.

Yes, all entrepreneurial ventures begin as small businesses. However, whether they have the potential to grow into large businesses as do other small businesses depends on a variety of interesting factors.

Most entrepreneurs are self-employed insofar as they work exclusively for themselves, manage their own time and resources and set goals, etc. However, self-employment for biomedical entrepreneurs is often more challenging than it is for other entrepreneurs. Among other things, biomedical entrepreneurs must deal with highly technical content, sophisticated equipment, huge development costs of their products, and the FDA codes and regulations that regulate the marketing of medical products.

Often, persons want to become entrepreneurs because they don't want to work for someone else. They view entrepreneurship as a way in which to gain some control over their lives, to become their own boss and to achieve a level of independence. Certainly, becoming an entrepreneur can brings all of these to some extent.

However, entrepreneurship also can be as constraining as it is liberating. For example, when you own and operate an entrepreneurial venture, you cannot just resign your job and change employment if you become frustrated, bored or discouraged. To succeed, you often must tough out the day-to-day grind, see difficult tasks through to completion, live with disappointing outcomes and find the stamina to try again tomorrow.

Entrepreneurs spend a lot of time thinking through issues. How can I develop a new product or significantly improve an old one? How can I create a new market or penetrate an existing one? How can I acquire new customers?

Sometimes, these questions are easier to answer than at other times. For example, if you are going to create something new, you must have the ability to determine, at least to some extent, how to develop the product and how to find or create a market for it. However, if you're going to charge ahead into an existing market, you must be absolutely certain that your device is so superior to that of your competitors that you can set the rules for dividing up the playing field. A product that offers only a minimal improvement over that of a competitor may limit you to playing by the rules that other people have already established.

It is very difficult to come out on top when you have to play by someone else's rules. This does not suggest that the rules in place are unfair. However, it is always preferable to create your own rules, to create your own game, and to define your own basis for competition than to buy into someone else's well-established game.

3.5. J. Craig Venter, Celera Genomics and Human Longitivity, Inc.

On June 20, 2000, U.S. President Clinton, UK Premier Blair, NIH Director Collin and President of Celera Genomics Venter, ambassadors and scientists met in the East room of the Whitehouse to celebrate the completion of the sequencing of the entire human genome. This had been accomplished by the consortium of the International Human Genome Project with a budget of $3 billion and Celera Genomics with an investment of $300 million. During this celebration, Clinton said"

We must sort through this trove of genomic data to identify every human gene. We must discover the function of these genes and their protein products, and then we must rapidly convert that knowledge into treatments that can lengthen and enrich lives."

Celera was founded in 1998 by J. Craig Venter and Applera. Separately, but in parallel with the International Genome Project, it sequenced a first draft of the human genome in 2000. The competition between the consortium and Celera had accelerated the genome sequencing program by years. Later, it was disclosed that one genome sequenced by Celera came from Venter himself. When the sequence was published, Venter was reported in a press briefing as the first human being to gaze upon his entire complement of DNA, all 6 billion chemical units. Venter left the firm in 2002, and eventually Celera was acquired by Quest in 2007 for $671 million.

In 2007, Venter wrote the book entitled *A Life Decoded* to tell the story of how he revealed the human genetic code by a combination of vision, tenacity, ingenuity, and insight that challenged conventional wisdom and refuted his critics (5). Growing up in California, Craig Venter was an unremarkable student, but with little interest in his schoolwork and even less motivation to complete his education. However, being drafted and serving as a Navy medic in Vietnam piqued his interest in science and medicine. Consequently, after jump-starting his education, he received advanced degrees and quickly established himself as a gifted and outspoken scientist. At the National Institute of Health he introduced novel techniques for rapid gene discovery and, at his own research institute in 1995, sequenced the first genome of a living species in history, the bacterium *Haemopbilus influnzae.*

Venter is regarded as one of the leading scientists of the 21st century for his numerous invaluable contributions to genomic research. While in NIH, he invented the expressed sequence tags method for rapid identification of all mRNA strands present in a cell. He is known as one of the first persons to sequence the human genome and the first to transfect a cell with a synthetic genome. Following his move to La

Jolla where UCSD is located, he founded or co-founded the following non-profit organization and corporations:

- J. Craig Venter Institute, a not-for-profit, research organization with approximately 250 scientists and staff that are dedicated to human, microbial, plant, synthetic and environmental genomic research, and the exploration of social and ethical issues in genomics.
- Synthetic Genomics Inc., a privately held company that concentrates on developing products and solutions using synthetic genomics technologies.
- Human Longevity Inc. (HLI), a San Diego-based genomics and cell therapy-based diagnostic and therapeutic company that was formed to extend the healthy, high performance human life span.

HLI intends to amass the world's largest database of human genetic and medical information in the hope that the information can tease out the roles of genes in all diseases, allow accurate predictions of people's health risks and suggest ways to avoid those problems (4). HLI also sells genetic workups for as little as $250. However, $25,000 gets you "a physical on steroids" that includes a full analysis of subjects' genomes, sequences their gut bacteria or microbiome, analyzes more than two thousand other body chemicals and put them through a full-body MRI scan. HLI aims to sequence a million genomes by 2020. To generate income, HLI will sell the information to insurance companies to better servicing their clients and to biotech companies to develop new medicines. It is too early to tell whether this business model will cause HLI to be profitable.

In an article on the entrepreneurship of Venter, Antonio Regalado, the senior editor of Biomedicine of *MIT Technology Review* had this comment (2): "*Venter has a history of selling big concepts to investors and then using their money to carry out exciting, but not necessarily profitable, science.*"

3.6. Career Entries to Entrepreneurship

The best entrepreneurs come in one of two forms. The first type of entrepreneur has a feeling and an affinity for people. He or she recognizes that there are many opportunities to exploit and seeks them out. The second type of entrepreneur is more technically oriented. He or she has a new or different way of solving a problem or making use of a platform (or enabling) technology. By solving a whole class of problems with a certain technical approach, this second type of entrepreneur seeks out ways of implementing this technical approach commercially.

Although business schools attract and graduate numerous entrepreneurs, they are not the largest source of entrepreneurs. Business schools provide extensive knowledge of running a business with great efficiency. Consequently, business schools are great places to find chief financial officers, marketing experts, operation managers and strategy planners for your entrepreneurial team, but not necessarily people who will take or manage risks.

The traits or skills that are necessary to become an entrepreneur are leadership, perseverance, passion and courage. These are attributes that are widely distributed throughout society. Schooling is not necessarily a requirement. Anyone who posses these attributes has the potential to become a successful entrepreneur, as long as he or she also has a great idea, a way to raise the operational capital, and a means to sell the product.

As an alternative to going to business school, several universities offer a Master of Engineering Management program for engineers who want to manage and operate a complex business in this new century. Courses are offered in marketing, management and finance and accounting of high tech companies. To enhance their management skills and careers, engineers can choose to participate in the program offered by Engineering Management Certification International of ASME.

Students can learn about the entrepreneurial game while they are still in school by participating in fund raising activities for their college and university, or by helping existing firms to market their product lines. Hands-on experiences like these are great ways to pick up basic skills and gain credibility within a specific organization — credibility that may lead to an invitation to continue full time after graduation.

An enterprising student can always find an alumnus, or alumna, who owns his or her own business and ask him or her to be put to work. Using such old-school ties is a great way to get one's foot in the door and gain industry experience.

Most cities have local technical councils or organizations that actively promote entrepreneurship and help people to develop and prepare business-plan presentations. Becoming involved with these councils or organizations is another good way for students to study new ventures or participate in a small three- or four-person technical team and learn how to write a business plan or business concept.

Another entry point to entrepreneurship is by gaining employment with a family business. If you have no prior business experience, look around for business people who know you, or who know someone you know. These are the people who are most likely to give you the greatest break. If you like their business, you could offer to work for them or volunteer your services for the summer, or part-time during a school term, or for a year or two after you graduate.

Students who want to gain some business experience can also seek out academic entrepreneurial ventures—new businesses that have been built on the research of one or more professors. These days, a growing number of professors have businesses on the side that are sanctioned and supported by the university at which the professor works. Most of these business activities are licensed by the university. However, very few professors have much extra time on their hands with which to fully develop their ventures into commercial successes. Consequently, they may welcome any additional help that they can get. Students specifically could look for opportunities to work on the development of a new or improved technology, while learning something about the business aspects of the venture at the same time.

Resourceful students may find other career entries to entrepreneurship. The important thing to remember is that there are many ways for would-be entrepreneurs to gain business experience without actually having to attend business school or to commit to working in a particular industry.

3.7. Practicing Entrepreneurship

What do you really need to start a venture? Your product must have customers who are willing to trade resources with you, whether or not these are cash, in exchange for value. This is the starting point of the business. If you have enough customers, you may not need to invest any personal cash, as the banks may be willing to lend you their money. On the other hand, if a great deal of research and development is required before you can attract customers, you will need to fully commit your personal resources and bear all of the risk up front.

A safe strategy for accomplishing this is to break the business development into manageable pieces and then do them one at a time. If the first piece works, do the second one. If that works, do the third! As mentioned at the beginning of the chapter, the successful entrepreneur is one who can determine how to carve the entire project up into digestible pieces and get each one done with minimum financial risk.

Many universities offer courses on innovation, design and entrepreneurship. Students who are interested in biotechnology and biomedical engineering could actually generate a venture idea from class activities.

However, it is important to note that not all courses in innovation, design, and entrepreneurship are created equal. Look for those in which the curriculum is not shaped around a very specific idea, but, instead, provides a range of topics in the coursework, so that you can follow your specific passion or focus of interest and still learn something relevant.

When do you want to begin your entrepreneurship? There is a temptation to put it off and to think of it as a long-term proposition — something to consider in 10 years, or maybe 15, when you're most likely to be in a position, professionally and financially, to do what you really want to do.

However, during those 10 to 15 years, you will probably marry, start a family, and buy a house. Before you know it, the freedom to do what you want will be gone.

Consider this observation of Professor Wendell Dunn. Years ago, Dunn was an emergency medic. He saw a lot of patients who, according to their families, died before ever accomplishing what they really had wanted to do. They had continued to defer their dreams for a variety of reasons, and had borne the pain of never reaching where they had wanted to go.

So when it comes to deciding when to start your entrepreneurial venture, we say, "Do it now!" At least, make the commitment and get started. If you put it off, it probably won't get done.

One way to think about beginning your entrepreneurship is for you to start with the industry. Pick a highly-promising industry, like biomedical engineering, and say, "I am going to learn about the structure, conduct, and people of this industry." Learn who the big players are and who the small players are. Learn who will fail and who will succeed, and why. Learn what the rules are by which these businesses compete.

Learn how well the winners perform in comparison to the losers. Learn who is making money and why patients want their products. Learn about the technical content of successful products by their practices in science, engineering and medicine. Finally, learn whether or not the industry is poised for extraordinary growth—so that you will know if you want to pursue your entrepreneurship in it.

If you do all of these things, you will find that biomedical engineering is a profession that is worthy of your entrepreneurial commitment and devotion. Biomedical engineering is *the* industry to pursue in this Medical Device and Biotech Century.

3.8. Development of Wearable and Wireless Devices

With the progress in electronics, smart phones, internet and cloud computing and genetic engineering, we summarize below the development effort make in the following two areas and their medical importance and the importance of these industries to healthcare:

Wearable Devices. The emergence of this industry can be illustrated by the sales of Apple Watch and the growth of Fitbit. First, the development of Apple watch was announced by Apple CEO Tim Cook in September 2014 as a device, which incorporated fitness tracking and health-oriented capabilities and was integrated with Apple mobile operating systems. When Apple Watch was released in April 2015, it quickly became the best-selling wearable device with sale of 4.2 million smartwatches in the second quarter of the 2015 fiscal year.

Seeing the potential of using small sensors in small wearable devices, James Park and Eric Friedman founded Fitbit in 2007 with an initial capital of $400,000. Fitbit launched its tracker at the end of 2009, shipping around 5,000 units with a further 20,000 orders on the books. Because the sales were made directly to customers, the company had a good profit margin. However, the two founders decided to raise more capital and to team up with retail chain Best Buy to sell their products. On the website Medtech.com, Barry Morrow described their success as a company with a circuit board in a wooden box in 2007 to a $4 billion company in 2015. Fitbit has sold more than 48 million wearable devices in 64 countries. Speaking to *TIME*, Park said that the "next big leap" would come when "we tie into more detailed clinical research" and create devices that can make "lightweight" medical diagnoses.

It may be also noted that the news that emerged in late 2016 included the following comments:

> *"Smartwatches may not be cool anymore. Worldwide sales plunged 52% in the third quarter of 2016 compared with the same period last year, according to a report by the market analysis company IDC."*
>
> *"The novelty for fitness trackers is not there anymore. Consumers are a very fickle bunch." Georgia Wells in the Wall Street Journal wrote, "About 30% of people in the U.S., U.K., and Australia who had ever worn a fitness tracker no longer use the device, according to a survey this summer by market-research firm Gartner Inc."*

In response, Fitbit is trying to reposition its products from "nice-to-have" to "need-to-have," according to Fitbit CEO Park, as quoted by Wells. The goal is to have doctors prescribe the devices to, for example, monitor patients' heart conditions.

WirelessMedical Devices (7). Today's Medical Development Magazine gave the following descriptions of this industry:

> *"Wireless medical devices simplify the ability of healthcare professionals to access information on the conditions of individuals receiving care while providing ease of mobility to patients. Wireless technologies are enabling medical practitioners to monitor and provide treatment remotely while patients are in hospitals, medical clinics, ambulances, or home."*
>
> *"The wireless medical device industry offers significant potential, with an expanding global market and huge demand for improved healthcare services driven by an increasingly older population and improved medical insurance coverage. The demand for wireless medical devices is expected to continue to increase in the coming years as innovations in sensor and wireless communications make these devices ever more capable of monitoring changes in the human body and facilitating treatment."*

The technology can be a new micro-implant that consists of sensors that can be programmed to monitor or adjust health conditions from inside the human body or a novel bio-electronic device known as electroceuticals, which can be used as substitutes for drugs to adjust bodily functions.

3.9. Entrepreneurship and Visions of Uncas A. Whitaker

Mr. Whitaker had built up the AMP Inc. to become the world's leading manufacturer of electrical, electronic, and fiber-optic connector and inter-connecting systems (1). In addition to his legacy of entrepreneurship, Whitaker and his wife, Helen, left their wealth to The Whitaker Foundation to promote the growth of a new engineering discipline — biomedical engineering (6). Whitaker is a true role model for all who work in biomedical engineering. His career path is summarized below to illustrate how he prepared himself in becoming an entrepreneur and a visionary.

Pursuit of Education and Career Development. Whitaker obtained his B.S. in Mechanical Engineering from MIT in 1923. He took courses in political economy in which MIT had earned a reputation for first-rate training, not only for engineers, but also for future business administrators. His thesis, "The Design of a Rubber Heel Testing Machine," gave him an opportunity to develop a practical approach to problem formulation, problem solving and, most importantly, a belief that an engineer needs to fully understand the requirements of his customers and then design the best machine in order to serve them well. This viewpoint is a major tenet of Whitaker's engineering and business philosophy in building his career and company.

He joined Westinghouse Air Brake to work as an assistant foreman in the engineering division. In this environment, he was encouraged to concentrate on original development. During his final two years with the Westinghouse Company, he filed some twenty-seven of his patentable ideas. In the meantime, he earned a B.S. in electrical engineering in five semesters from the night school of the Carnegie Institute of Technology (now Carnegie-Mellon University). The attraction of a job that combined engineering development and management responsibility led Whitaker to accept the position of division head of Development and Design in the Hoover Company. One of the major challenges that Whitaker took on was the design and production of an electric cleaner that could be offered for sale at a price that people during the depression era would buy. Although this had been tried without success, Whitaker developed in five years the Model 150 vacuum cleaner that had a selling price of $79.50. This was 30% lower than the price of the Hoover Company's previous model and credit terms of $1.50 weekly, payable monthly, made this new vacuum cleaner a hot seller. In the nine years (a depression period) of his employment at Hoover, the patents that were filed under his name and the doubling of his salary from $3,000 to $6,441 from 1929 to 1938 attest to his remarkable accomplishments. During this period, Whitaker also completed law school and passed the Ohio bar examination in 1935.

American Machine & Foundry Company was looking for an engineering business type of person to reorganize its entire engineering function and decided that Whitaker was the man that it needed. As Director of Research and Standards and having assignments of a rather broad nature, he described his job as:

- Reorganization of the design and cost-estimating activities so that the cost of any particular machine can be established before design work has started, rather than after the design has been completed. The reduction of manufacturing costs will be part of this assignment.
- A study of all technical activities of the company with a view to working out a long-term plan for the future development of the engineering and other technical divisions.
- A study of the patent activities of the company in cooperation with the Patent Department.

This job enabled him to gain valuable experience and insights in decision making at the highest corporate level. He built a reputation as not merely a successful engineer, but as a brilliant businessman. During the interview to fill an executive position, he so impressed the important business management firm of Booz, Fry, Allen and Hamilton that it produced a future opportunity for Whitaker to run his own "little firm" that he could build into something big. His theory of building a product was to out-engineer all others. This led to the semi-official slogan, "Engineering the hell out of a product," of AMP Inc., which he founded in 1941.

His friend and associate, George Hastings, described Whitaker as always being profit-minded. He did not just want to design things; he wanted to design things for a profit. The people who came to work for him paid a lot of attention to what things cost. Hastings further stated that Whitaker was always able to develop a good deal of loyalty in his engineers. When he had his own company, many of them were determined to go with him.

Acquisition of AMP. In the fall of 1941, Whitaker along with a handful of associates and investors took over a small firm that was producing solderless electrical terminals for the aircraft and marine industries. Whitaker, as Vice President and General Manager of AMP, oversaw the activities of six divisions: Financial, Sales, Engineering, Patent and Legal, Production, and Inspection. He also headed the Engineering Division, which was composed of three departments – Production Engineering, Experimental Engineering, and Research. In the first three years, Whitaker discovered that solderless electrical terminals, despite their excellence in concept and saving in labor cost, did not sell themselves. It was a formidable task to convince the aircraft and marine industries to switch from the well-established practice of connecting wires by twisting and soldering to the more cost effective method of using solderless connections. Fortunately, Whitaker was able to reorganize the company by replacing the previous owner of the firm with G. Earle Walker, who built a powerful and successful sales organization for the firm. With patience and a lot of hard work, the Sales Division gained such important new accounts with familiar names as IBM and Douglas Aircraft. With the introduction of the Diamond Grip Insulation Support Terminal, which immediately became a best-seller in mid-1942 because of its superior electrical and mechanical properties and its lightweight and compact form, AMP developed into a real factor in the competitive field of solderless connectors.

Whitaker always made a point of finding the best men and then giving them a free hand to carry out their responsibilities. This policy applied not only to key management personnel and engineers, but also to those who sat on the company's board of directors. Many of these persons were former associates of Whitaker at Westinghouse Air Brake, Hoover, and American Machine & Foundry Company. Whitaker understood the need to be particularly careful in leading his engineering talents, both in protecting them from distractions and in managing their activities to mitigate frustration. He knew that creativity had to be nurtured, as well as watched over.

He had frequent patent meetings with engineers during which he asked them what they had accomplished. In Whitaker's view, the primary purpose of patents was to create the spirit of competition in his engineers and to spur them on to "engineering the hell out of a product." In these meetings, Whitaker not only was an outstanding source of encouragement, but also the boss who put the emphasis on finding ingenious solutions to problems. This not only made solutions patentable, but also provided a great commercial advantage.

In 1940s, the annual sales of AMP grew from \$2 million to \$3 million. The average profit after tax was in the range of 1 to 4% of sales. However, the next decade would prove that Whitaker's optimism about the electrical termination business was correct. In advance of many other American industries, Whitaker led AMP into international markets with the establishment of many wholly-owned subsidiaries overseas. As a result of engineering development and sale efforts, Whitaker built the company into a Fortune 500 company by 1965. Less than 25 years after its founding, AMP had become a multinational enterprise with subsidiaries all over the world and annual sales well in excess of \$100 million. In the 1960s, AMP experienced an annual growth rate of 20%, twice the growth rate of the entire electronics industry. By now, AMP had grown to become the world's leading manufacturer of electrical, electronic and fiber-optic connectors and interconnection systems with annual sales in the range of \$10 billion.

Successes and Growth. In receiving the 1960 Growth Company Award from the National Association of Investment Clubs, Whitaker stated the following:

"There are two basic reasons for AMP's continued success and growth – accomplished, incidentally, without benefit of mergers or acquisitions. The first of these reasons – the engineering approach, the second – our direct sales method.

At the outset, AMP adopted a very broad engineering approach and applied it to a confined product area. This approach was applied to the wire termination field and to this day is very appropriately described as "Precision Engineering Applied to the End of a Wire." This field has been the backbone of the company, our bread and butter items, and will continue to be so during the next few years. As a result of this precision engineering, AMP produces complete terminations that usually reduce termination costs through speed, simplicity and uniform quality of application.

An engineering approach such as this is no small task as evidenced by our present customer list, which numbers in the thousands, by our 15,000 product items, and by some 2,000 patents issued both here and abroad. It requires spending some 13 or 14% of our sale dollars to maintain a staff of over 500 persons in research, development, and engineering activities, the majority of whom are graduate scientists or graduate engineers or engineers specially trained in the areas of physics, electronics, chemistry, electro-mechanics, atomic energy or other related fields.

The foundation of this (direct selling) program is a force of highly trained District Sales Engineers exceptionally qualified to work with the customer's own engineering and other technical personnel. At staff level, we have both industry sales sections concentrating on individual industries and their changing technology and product sales managers promoting our newer lines of products."

Whitaker believed in competition and the associated notion that the business that did things better than its competitors would survive. Whitaker expressed this business philosophy in the following statement:

"The success of a business venture depends much more on the intelligence, ability and drive of its employees than on low taxes, low interest rates, or low labor rates. For instance, the wages in our plants in the United States are the highest of any in the world, yet we produce most of our products at a lower cost than anywhere else."

Legacy of Mr. Whitaker. All these entrepreneurial accomplishments and his endowment to, and vision for, the biomedical engineering discipline will always be remembered by us. The contribution to healthcare that he envisioned for biomedical engineers is best summarized by the following excerpt, which was written by Ruth Whitaker Holms, Ph.D., Foundation Committee Chair, Whitaker Foundation in 1992: (6)

"My father Uncas A. Whitaker had two distinct visions, and he pursued both during his lifetime. The first involved the creation of an engineering-based manufacturing company, and he lived to see AMP Incorporated become the world's leading producer of electrical connecting devices. His second vision was to expand significantly the use of engineering techniques and principles in biomedical research. During his lifetime, he initiated and supported biomedical research programs that encouraged the interaction of engineers with scientists and physicians. Biomedical engineering research, however, was only in its infancy when he died in 1975. The full development of biomedical engineering as a research discipline is the mission of The Whitaker Foundation.

After my father's death, a biography was written of his life with the title 'The End Is Just the Beginning.' The reference is to the end of a wire where the AMP connectors were attached and also to the future of his company after his death. It is our hope that the title also will appropriately describe the conclusion of The Whitaker Foundation in 2006. If the Foundation programs are successful, biomedical engineering programs will have superbly trained faculty investigators and the supportive environment needed to make even greater contributions to the prevention and treatment of disease and injury. When this occurs, my father's vision will be realized, and The Whitaker Foundation will have completed its mission."

References

1. Cohn, W. H., *The End Is Just the Beginning*, Carnegie Mellon Univ., 1980.
2. Regalado, A., J. Craig Venter to Offer DNA Data to Consumers, September 22, 2015; www.technologyreview.com/s/541516/j-craig-venter-to-sell-dna-data-to-consumers/
3. Rodengen, J. L., *The Ship in the Balloon*, Write Stuff Enterprise, 2001.
4. Rojahn, S. Y., Microbes and Metabolites Fuel an Ambitious Aging Project, MIT Technology Review, March 11, 2014.
5. Venter, J. C., *A Life Decoded*, Penguin Group, 2007.
6. The Whitaker Foundation, 1992 Report.
7. www.todaysmedicaldevelopments.com/article/wireless-medical-device-industry-forecast-2020-113016/

Chapter 4

Establishing Your Venture Invention

4.1. How an Entrepreneur Acquires a Venture Idea

The success of Whitaker in building AMP Inc. (described in Section 3.9) illustrates that one way to acquire a venture idea is by acquisition. However, before you do that, you will need to have built up a reputation as someone who can run a successful business and knows the market potential of your venture. Both are important when seeking venture-capital or a bank loan for your acquisition. An example of the importance of the team in an acquisition is provided by Abele and Nicholas of Boston Scientific (Section 3.2). Finally, to put your business on a growing path as Whitaker did, you must follow his lead and hire outstanding engineers and business professionals — perhaps people whom you already know through established networks or previous contacts — and make a firm investment in engineering innovation, marketing and sales.

As a middle ground between the de novo creation of a new entity or the acquisition of an existing business, there is the establishment of a partnership with other like-minded entrepreneurs. For an example, we can look at Adenosine Therapeutics, LLC (ATL), which was founded by Robert Capon and Joel Linden in 1999.

Capon earned his Bachelor of Science degree in mechanical engineering from the University of Virginia and then worked in marketing for Hewlett-Packard for six years. He next earned a Master of Business Administration degree from Harvard University and then worked at General Electric (GE) for three years, eventually serving as manager of a venture development team within GE. After his tenure at GE, he successfully built up three biotech companies. Two were sold to private owners and one, a pioneer in antiviral drug resistance testing known as Virologic, successfully completed its Initial Public Offering (IPO) and is now traded on the NASDAQ exchange.

Meanwhile, Linden was working as a professor of medicine at the University of Virginia. His research interest was cardiovascular physiology. Over the years, Linden worked with a number of colleagues to successfully construct a series of novel synthetic adenosine derivatives.[a] With this success behind him, Linden began looking for a partner with whom to create a company that would investigate the therapeutic potential of the derivatives in treating inflammation, heart disease, asthma, chronic obstructive pulmonary disease (COPD), diabetes and Parkinson's, and then commercialize the adenosine derivatives for use in therapy.

In the late 1990s, Capon made it known to the technology community that he was looking for a new biotechnology company to develop. After a period of investigation and mutual due diligence, Linden concluded that Capon was, in fact, a world-class business person. Capon verified that Linden was a well-established researcher in adenosine and the two joined forces to form ATL. With these two exceptional individuals at the helm, Bristol-Myers Squibb, a global pharmaceutical company, had sufficient confidence in ATL to offer it a fee for the right to license the firm's adenosine derivatives for the treatment of digestive disease. This funding added capital to support the company's operation. The combination of Capon's business savvy and Linden's research excellence also enabled ATL to win numerous Small Business Innovation Research (SBIR) grants to carry out the research necessary to achieve the company's R&D goals. Currently, the company works with approximately twenty laboratories on the clinical evaluation of adenosine derivatives.

Ventures can be developed not only as a result of acquisitions and partnerships, but also from ideas generated from your education or work experience. There are unlimited opportunities for the cross pollination of ideas in discussions with colleagues in different fields of expertise. Once the concept for a new business idea has been identified, it is important to solicit feedback by means of your connections with physicians, biomedical engineers, and other health professionals. After the idea has been initially validated, you can use your expertise — the knowledge and

[a] Adenosine is a naturally occurring substance in the body. It is produced by cells under stress. Linden's lab had used advanced technologies to identify the adenosine-specific receptors of cells that are responsible for regulating inflammation. The synthetic adenosine derivatives were then developed to determine whether these receptors could be turned off in order to turn down (or even reverse) the inflammatory responses of diseases.

skills that you acquired at the university you attended and at the places where you have worked — to create a biomedical-engineering solution.

Most new business ideas arise in one of two forms. The first form consists of newly developed technologies or processes that are applied to existing problems. The second form consists of an existing problem that lacks an acceptable solution or creative individuals who are looking for new and creative solutions. Many of the venture ideas that were responsible for the successes of Medtronic Inc., a medical technology company that was founded in 1949, fall into this "from problem to solution" category. The following section, which was condensed from *One Man's Full Life,* the autobiography of Earl E. Bakken (1), describes three Medtronic successes. The first involves the development of an external pacemaker as a venture idea that arose from the "blue baby" problem. The second concerns the opportunity that led to the invention of the Hunter-Roth electrode, which expanded the use of the pacemaker to Stokes-Adam disease. The third success entails the acquisition of someone else's venture idea by a licensing arrangement that gave Medtronic the right to commercialize the Chardack-Greatbatch implantable pacemaker.

4.2. Three Inventions of Medtronic, Inc. around 1960s

When Bakken, co-founder of Medtronic Inc., graduated from high school, he wanted to work with electricity. Because of films like *Frankenstein* — the unforgettable story of the learned doctor who, through the power of electricity, brings to life a man who was assembled from the body parts of corpses — Bakken became intrigued with the idea of using electricity to restore life. From a very young age, Bakken had read *Popular Mechanics* and *Popular Science* magazine voraciously. Despite the excellent instruction that he received at school, he considered many of the most important lessons that he had learned to have been self-taught.

After working as a radio instructor in the U.S. Army for three years, Bakken returned to Minneapolis to earn a BS degree in electrical engineering from the University of Minnesota's Institute of Technology. During this period and later, while working on his master's degree there, he became acquainted with several people who worked at the University Hospital, which was located across Washington Avenue from the

Institute. He provided them, on an ad hoc basis, with an on-the-spot repair service for malfunctioning medical equipment.

While at a family birthday party, Baker talked to Palmer J. Hermundslie, his brother-in-law, about the occasional work that he was doing at the hospital. A few days later, on April 29, 1949, Bakken and Hermundslie formed a partnership to service medical electronic equipment. This gave birth to Medtronic Inc. The two partners set up shop in a garage in Minneapolis. As told by Bakken, they grossed a grand total of $8 (about $72 in 2008 dollars) for the repair of a centrifuge during their first month in business.

They supplemented their modest repair and equipment-building activities for doctors in the following year by selling medical equipment for other companies. The increased activities not only kept them in business, but also provided an entrée to the labs and offices of doctors, nurses, and technicians whose needs, knowledge, and encouragement would prove to be critical to Medtronic's growth and success.

The "Blue Baby Problem". In 1954, Dr. C. Walton Lillehei of the University of Minnesota Hospital began to operate on infants who had congenital heart disorders that robbed their blood of oxygen and caused a bluish, or "cyanotic" cast to their skin, which gave rise to the use of the term "blue babies" to describe them. The operation, although effective, often interfered with the heart's ability to conduct electrical impulses to produce a regular heartbeat. This resulted in a condition known as "heart blocks."

To remedy this problem, Lillehei used AC-powered external pacemakers to stimulate the tiny hearts to beat following surgery. Built on vacuum-tube technology, the pacemaker was bulky, but could be wheeled around and plugged into a wall socket. On October 31st, 1957, a three-hour electrical-power outage in Minneapolis threatened the lives of the babies who had undergone surgery and depended on use of the pacemaker. Tragically, one baby died that night. The next day, Lillehei asked Bakken if Medtronic could come up with something better.

Bakken's first approach was to convert the DC voltage of an automobile battery to AC, and to use the battery to run the pacemaker. This was quickly deemed an inefficient way to resolve the problem, since only a 10-volt pulse was required to stimulate a baby's heart.

Bakken then recalled having seen a circuit for an electronic transistorized metronome in *Popular Electronics* magazine that transmitted clicks through a loudspeaker with an adjustable rate that adapted to the music. He dug up the back issue and built that circuit in a four-inch square, one-and-a-half-inch thick metal box with terminals for connection to the battery and wires to carry the pulses to the heart. In four weeks, he designed, constructed and tested in his garage shop the first small, self-contained, battery-powered, transistorized, external pacemaker that could be taped to a patient's chest wall. When pacing was no longer needed, the wires could be withdrawn without having to reopen the chest.

Bakken took the device to the university's animal lab and tested it on a dog. "Of course, it works," Bakken remembers having stated on completion of his test. The next day, as Bakken returned to the hospital to work on another project, he spotted a little girl who was wearing the prototype that he had delivered only the day before. Lillehei explained that his lab had told him that the pacemaker worked and that he had wanted to use the best technology available to save the child's life.

Bakken later wrote, "We [Medtronic] were struggling to survive. When a customer asked for something, we provided it as quickly as we could — and only then hoped it might somehow earn us a few bucks. We were operating — not by design, but by necessity — according to a principle that has since been popularized by the statement: *Ready, fire, aim!* We got the order, built the product, and only after rushing it to the customer, debated its long-term possibilities for the company. As it turned out, of course, that battery-powered pacemaker put us in the cardiac pacing business."

Because of the response of the medical community to the success of Lillehei's operations on blue babies, the general press soon picked up the story of the little Medtronic box that kept children's hearts beating after surgery. People were calling it a "miracle." By the end of 1958, Medtronic had received 60 orders for the pacemaker from all over the country. (To put this figure and the company's subsequent success with this product into perspective, consider the fact that, in 2003, a Medtronic-manufactured pacemaker or electrical stimulating device was being implanted into a patient every seven seconds!)

The experience and knowledge of Bakken, as well as his network of physicians, made Medtronic "*Ready*" in coming up with a solution to the problem of cardiac pacing. The company "*Fired*" the prototypical device in four weeks with the enthusiastic support of Lillehei. However, the company immediately found an imperfection in the design of the prototype pacemaker — the children on whom it was used thought that it was a toy and pulled at its knobs and switches. The model was immediately modified with the knobs and switches recessed, reflecting the "*Aiming*" part of the popular statement that Bakken had quoted. In other words, once an idea is ready, and the prototype is fired, the successful entrepreneur will always fine-tune his or her product so that its aim is still right on target. Since that time, the proven wisdom of "Ready, fire, aim!" has inspired Medtronic to continuously improve the functionalities of its products.

Treating Stokes-Adams Disease. The confluence of transistors, batteries, materials, and cardiac surgery created a fast-moving stream of new therapeutic possibilities. In 1959, Dr. Samuel Hunter of St. Joseph's Hospital in St. Paul, Minnesota, decided that Medtronic's battery-powered pacemakers might be used to good effect on patients who had Stokes-Adams disease, a chronic, progressive condition that results in long-term heart blockage in older patients. Consequently, Hunter and Norman Roth, a Medtronic engineer, developed a new type of bipolar electrode that could be sutured to the patient's heart to more effectively concentrate the pacemaker's current at the location where it was needed. This concentration enabled the current to be reduced by 70% with the electrode technology that was available. Soon, they had good clinical results, as demonstrated by a 72-year-old patient who lived for another seven years with the help of his Hunter-Roth electrode and pacemaker.

Implantable Pacemakers. Dr. William Chardack, a heart surgeon, and Wilson Greatbatch, an engineer, began the development of an implantable pacemaker in late 1950. This pacemaker was powered by batteries that had been newly developed by Greatbatch to work inside the human body for a long period of time. They carried out testing of the implantable pacemaker with animal studies. By early 1960, they had obtained encouraging results on the performance of this pacemaker when implanted in a human patient.

Chardack and Greatbatch became acquainted with Bakken after they expressed their interest to use the Hunter-Roth electrode for their implantable pacemaker. Recognizing the scientific and commercial opportunities of collaboration, Bakken signed a licensing agreement with them that gave Medtronic the exclusive right to manufacture and market the patented Chardack-Greatbatch implantable pacemakers in exchange for a royalty of 10% of sales. By the end of 1960, Medtronic's production department in Minneapolis had received 50 orders for this new device at $375 each, the equivalent of about $2,725 in 2008 dollars.

These three examples of Medtronic's success reflect the importance of networking in entrepreneurship. As Bakken stated, "*I truly believe that those personal connections have been at least as important to our success as our products. To my eyes, high tech has always depended on high touch to be effective.*" The image of a happy Bakken and Koch on their first meeting that appears in Fig. 4.1 clearly illustrates one successful result of personal connections and a high touch.

Fig. 4.1. A 1986 Florida meeting of Earl Bakken and a happy 5-year-old Lyla J. Koch who had received a Medtronic pacemaker when she was only a month old.

The Medtronic examples also illustrate the importance of building up your reputation as someone who can provide engineering solutions to medical problems and one who will diversify his or her product line when the opportunity to acquire an exclusive license of appropriate patented technology becomes available.

4.3. Evaluating Your Invention (Plan Selection)

Each step in the creation of your new venture raises the cost and commitment required to move to the next step. Therefore, it is prudent to evaluate any new business idea against a standard battery of marketability questions. Ten questions, grouped in three categories, are listed next for your consideration:

Patient Needs

1. Do physicians and patients want your device? It is important to meet the needs of customers, meaning those physicians, patients, biomedical engineers or health professionals who are most likely to use your product. Engineering a device and looking for its medical applications will not, in most cases, produce a product that can precisely fit the needs of the physician and patient. After a device has been developed, these customers should be continuously solicited for feedback on ways in which to improve your product.
2. Does your device solve an important medical problem? The importance of the medical problem may be closely related to the size of the market and patient demand for your device. Inventing a device or developing a new molecule that can improve or treat many clinical conditions or diseases may imply that you have a potentially large market for your product.

Superiority of Functions or Competitive Advantages. Intertwined closely with patients' needs, this category of superior functions will help physicians to improve their ability to care for their patients and enhance patients' quality of life.

3. What technological advantage does your invention have over that of its competitors? In the case of device development, an improvement in sensitivity by a factor of 10 will be extremely desirable for you to successively market your device. Also, you will have an overwhelming edge if your device is non-invasive, while the products of your competitors are invasive. Better performance of the device and robustness that make the device easier to operate by health practioners are two other factors that will add to the superior functions of your invention.
4. Do you have the capital and technology to build the device or to develop the drug? The more complex your device or drug is, the more capital you will need for its research and development. Whether the activity is R&D, clinical trials or marketing, it will takes time to complete. In many instances, R&D and clinical trials may yield a conclusion that the drug or device is ineffective, and that you

will need to begin again. Because of the extensive effort involved, and the risk of failure, the conventional wisdom is that between $300 million and $1 billion is required to produce an effective drug that becomes widely used. To be successful, an entrepreneur must develop a realistic estimate of the time required to complete the technology development in order to enable the company to establish a timetable for raising the capital to complete the technology and proceeding to the manufacturing and marketing phases.

5. Is the timing right? Bakken's success with his portable external pacemaker provides an example of good timing. First, transistors had already been invented. The transistors enabled Bakken to miniaturize the pacemaker that previously had been constructed with vacuum-tube technology. Secondly, open-heart surgery had progressed to the point where it created an opportunity for Bakken to develop a pacing device that would help to restore the heart's functions to normal.
6. What is your long-term plan to enable you to remain competitive and diversify your product line? In this high-tech era, you need a plan to continuously improve all aspects of your device to ensure that you can maintain your company's viability. Make sure that your company can produce more than one successful product. For example, the value of ATL, the bio-tech startup that was described earlier in this chapter, is high because the adenosine derivatives that it produces have potential therapeutic values for many diseases.

Overall Costs and Benefits. The increase in healthcare expenditures has been blamed on the introduction of expensive medical technology. In such a climate, the overall costs and benefits of introducing your device must be addressed.

7. How will the FDA regulate your device? In the opinion of many, working out a favorable FDA classification is a task that requires your utmost attention. The topics that are covered in Chapters 8 and 9 will help you to get a feel for the complexity and uncertainty involved in navigating FDA rules and regulations and how these rules and regulations will dictate the level of work that is necessary to get your product to market. Suffice to say that the more justification for your product that the FDA demands, the more capital you will need to get your device or drug to the marketplace, if ever. Therefore, this question is placed in the cost category.

8. Is your price competitive? In some cases, the price of the medical device is of little concern once it is regarded as cost effective by insurance companies and the government. In all cases, the price is much less important than the device's clinical power and its quality and your service. As evidence-based pricing becomes a requirement, you need to conduct a comparative cost analysis of your competitors' products so that health insurance will pay for the use of your device or product.
9. Will your device qualify for insurance reimbursement? If your competitor's devices are already in use with a medical procedure that has been approved by the Center for Medicare and Medicaid Services and the American Medical Association (AMA), your new device can be considered to be an improved use so that the existing reimbursement will apply. If the procedure that uses your device is new, you need to apply for its listing and reimbursement in a time consuming and expensive process that is described in Section 17.1.
10. What are the overall cost and benefits of your device to patients and hospitals? The overall cost of your device includes the cost to purchase the device and the cost of additional labor in the hospital to use it. Use of the device leads to a shortening of hospital stays, a shorter uses of the intensive care unit, a faster recovery outside of the hospital, a better quality of life of patients and an earlier recovery of the patients' productivity. It is still uncertain how these benefits will be considered by the government and public as part of the reduction of the overall cost. New methodologies to assess the effectiveness of medical therapies and technologies are on the way as a result of the initiatives of Health Technology Assessment and Value Based Purchasing that are described in Section 17.1. The costs and benefits of the device's use can be a factor for you to examine in order to minimize the impact on the marketability of your device or product.

For each of the companies that we examined previously, the answers to the questions above are definitely affirmative. Thus, each of these companies created a great product that benefitted patients, the company's reputation and the company's profit, as illustrated in Fig. 4.2.

A few more examples are cited later to further elaborate on the answers to these ten questions.

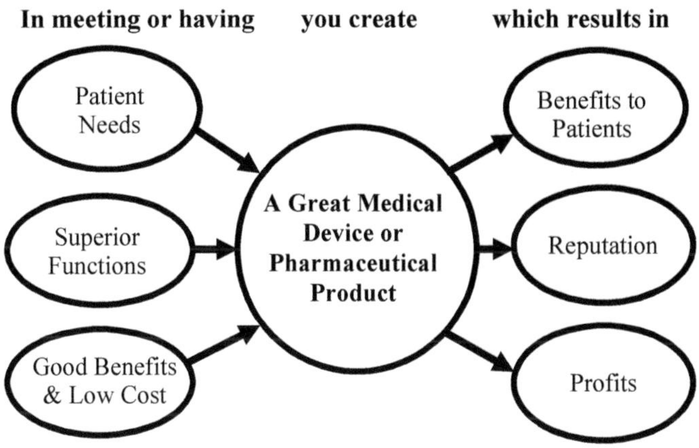

Fig. 4.2. Three key factors to generate a great product that provides benefits to patients and increases your company's reputation and profits.

4.4. Wallace Coulter's Invention, Venture and Legacy

Wallace Henry Coulter was an engineer, inventor, entrepreneur and visionary. He was co-founder and Chairman of Coulter® Corporation, a worldwide medical diagnostics company that is headquartered in Miami, Florida. The two great passions of his life were applying engineering principles to scientific research and embracing the diversity of world cultures. Most of the materials that are described here can be found on the website of the Wallace H. Coulter Foundation (6), including the first patent that Coulter submitted on August 27, 1949 and USPTO issued as patent number 2,656,508 on October 20, 1953.

An Elegant Idea Became a Company. *In 1935, he joined General Electric X-Ray as a sales and service engineer in the Chicago area where he serviced medical equipment. Later, he was transferred to the GE office in Manila to make sales and service calls to the entire Far East. After the war, Wallace returned and worked for several electronics companies, including Raytheon and Mittleman Electronics in Chicago. He maintained a laboratory at home to work on promising ideas and projects. One such project was for the Department of Naval Research. In this project, Coulter was trying to standardize the size of solid particles in the paint that was used on US battleships in order to improve its adhesion to the hull.*

He began tinkering in his garage laboratory in his spare time, experimenting with different applications of optics and electronics. Upon returning to the garage on one cold, blustery evening, Wallace was faced with a challenge. The supply of paint for the experiment had frozen while he was out. Not wanting to go back out into the cold, he asked himself, "What substance has a viscosity similar to paint and is readily available?" Using his own blood, a needle and some cellophane, the principle of using electronic impedance to count and size microscopic particles suspended in a fluid was invented - the Coulter Principle.

Remembering his visits to hospitals, where he observed lab workers hunched over microscopes and manually counting blood cells that were smeared on glass, Coulter focused the first application on counting red blood cells. This instrument became known as the Coulter Counter™. This simple device increased the sample size of the blood test by 100 times that of the usual microscope method, counting in excess of 6,000 cells per second. In additional, it decreased the time it took to quantify cell counts of blood from 30 minutes to fifteen seconds and reduced the error by a factor of approximately 10.

US Patent 2,656,508. The following quotation from his 1953 patent describes how and why he invented the Coulter Counter.

"This invention relates to means for detecting and counting particles suspended in a fluid medium.

In the counting of particles of the type contemplated by my invention, heretofore it has been difficult to obtain accurate results because of the limitations inherent in methods used and in the associated apparatus required. Some previous methods have been based upon colorimetry, which is the utilization of color density of a given sample to obtain a measure of the number of particles suspended, when compared with other known standards. In certain other methods, the sample is placed in a counting chamber and the particles are actually counted under a microscope, or the sample may be placed between microscope slides for counting.

Information of the type obtainable by determining particles concentration is especially valuable to the medical profession. Blood cell counts as well as the counts of other particles in the body fluids serve for many diagnostic purposes and as well to evidence of physical and pathological conditions of different parts and organs of the body.

Obviously such information, in order to be of great value, must be obtained quickly and must also be accurate. The previous methods alluded to were not only subject to great inaccuracies and required considerable time to be accomplished properly, but as well required the services of highly skilled technicians and hence were not readily available to all who required the same.

A further object of my invention is to provide new means for detecting and counting particles whereby a sample of the concentration to be measured is caused to pass through a constricted path and the presence or absence of a particle in the constriction gives rise to a detectable change in the electrical characteristics of the path."

Early Developments. *That same year, two prototypes were sent to the National Institutes of Health for evaluation. Shortly after, NIH published its findings in two key papers, citing improved accuracy and convenience of the Coulter method of counting blood cells. That same year, Wallace publicly disclosed his invention in his one and only technical paper at the National Electronics Conference, "High Speed Automatic Blood Cell Counter and Cell Size Analyzer."*

In 1958, Wallace and his brother, Joseph Coulter, Jr., founded Coulter Electronics to manufacture, market and distribute their Coulter Counters. From the beginning, this was a family company, with Joseph, Sr. serving as secretary-treasurer. Wallace and Joe, Jr. built the early models, loaded them into their cars and personally sold each unit. In 1959, to protect the patent rights in Europe, subsidiaries in the United Kingdom and France were established. The Coulter brothers relocated their growing company to the Miami area in 1961, where they remained for the rest of their lives.

Coulter Corporation. *During his tenure as chairman, the company developed into the industry leader in blood cell analysis equipment, employing almost 6,000 persons, with more than 50,000 instrument installations. The company has spawned entire families of instruments, reagents and controls not just in hematology, but also in flow cytometry, industrial fine particle analysis, and other laboratory diagnostics.*

Coulter insisted that the company control the entire supply chain: research and development, manufacturing, distribution, sales, financing, training and after-market service and support. In fact, Wallace's vision is illustrated by the fact that the Coulter Corporation was the only diagnostics company to support its instrumentation with the required reagents and quality control materials to operate the full system. His own personal field experience in sales and service helped in formulating Coulter's "Rule of a good salesman — List the positives and concentrate on them and after you make the sale, service the needs of your customer the best you can. Do this and you will build a loyal customer base that will stay with you in the hard times."

Landmarks in hematology. The following appeared in the article "*Fifty years of Hematology Innovations*" written by Berend Houwen MD, Ph.D., Medical Director of Beckman Coulter Inc.

"The Coulter Principle—and Coulter Counter—were landmarks in the field of hematology," says John A. Koepke, MD, Professor Emeritus at Duke University in Durham, NC, and one of the first to purchase the Coulter Counter Model A when he worked for the University of Kentucky Medical Center in Lexington, KY.

Before this technology existed, lab technologists had no choice but to do all their blood counts by hand. The Coulter Counter literally brought automation to hematology labs for the first time."

The Coulter Counter "essentially changed the face of medicine by turning hematology guesswork into accuracy," Dr. Jones says.

"While the previous technology provided a standard precision deviation of ±10%, the Coulter Counter reduced this deviation to ±1%," says Jones. "This made a significant difference in clinicians' ability to make accurate diagnoses and, more importantly, exact measurements of change in patients undergoing treatment."

Initially, the original Coulter Counter helped clinicians count and measure only red blood cells. Later, as the technology and instrumentation evolved, it enabled clinicians to also count and measure white blood cells. In the 1970s, it progressed again, enabling technologists to separate platelets—particles that

are less than one-tenth the size of red blood cells. The Coulters thus demonstrated that rare hybrid of business foresight and scientific exploration in developing their products.

The Legacy of Coulter. His deepest passion was to improve healthcare and make the improvements available and affordable to every one. He dedicated his wealth to the Wallace H. Coulter Foundation to continue the improvement of healthcare through medical research and engineering. The Foundation has funded translational research partnerships of many universities to promote collaboration between biomedical engineers and clinicians, to support the movement of promising technologies to clinical application, and to develop sustainable processes. The ultimate goal of these partnerships is to generate outcomes that will save, extend, and improve the lives of patients, who suffer from any disease or condition, in any size of market, in any discipline, in any country throughout the world. This goal of the Foundation reflects Coulter's values, which are endless curiosity, continuous learning, teamwork and consideration and respect for the individual, ethics and integrity.

4.5. The Venture of Howard Diamond

After receiving his doctoral degree in physics and electrical engineering, Howard Diamond became a professor of electrical engineering at the University of Michigan. Wishing to concentrate on the development of a venture idea, he left his university position after 10 years and formed Transidyne General Corporation to build and market his patented microelectrode technology as chemical sensors.

While exhibiting at an instrumentation convention, Diamond met a physician who inquired about the use of the company's technology to monitor blood parameters for open-heart surgery. At that time, blood samples were collected and delivered to a clinical chemistry laboratory for cumbersome and on-line, non-real-time measurements. Some of the measurements were being made with Diamond's microelectrode systems in the lab.

Although various people had tried to place an array of sensors into an extra-corporeal circuit to monitor blood parameters during open-heart surgery in real time, the idea had never been successfully marketed for two key reasons. First, the FDA considers direct contact with a patient's blood as invasive. Therefore, the FDA had required that extensive clinical studies be conducted to demonstrate the safety and effectiveness of the sensor array. Secondly, because sensor outputs drift over time, they must be calibrated in real time — an impossible task with the sensors in situ.

Diamond came up with an ingenious design that avoided the invasive classification and the calibration problem. His utility patent (Number 4,798,655: "Multiparameter Analytical Electrode Structure and Method of Measurement") was obtained in 1990, and a new company, Diamond Electro-Tech Inc. was founded to develop the blood-gas monitor for use in the emergency room and for critical surgery in the operating room.

Diamond's design included an array of sensors that provide a surgeon with real-time measurements of oxygen tension PO_2, PCO_2, pH, hematocrit, and concentration of K^+ and Ca^{++} of blood from the venous line (i.e., blood coming from the patient's vein), as well as those measurements of blood flowing to the arterial line (i.e., blood that was oxygenated for return to the artery). As designed, a minute amount of blood was withdrawn to flow past a micro chamber that contained an array of six blood gas and electrolyte sensors. The filling volume of the chamber was about 80 μL, and the blood was discarded after its passage through the sensors. Then, a standard calibration solution was switched on and flowed past the sensors for calibration. Because only blood that was withdrawn from an extra-corporeal circuit was used and then discarded, the FDA considered the monitor to be non-invasive and approved it within two years as a Class II device (see Chapter 8 for more information on FDA device classifications).

In 1988, Diamond Electro-Tech was sold to Mallinckrodt, a global manufacturer and marketer of healthcare products in the fields of respiratory care, diagnostic imaging, and analgesic pharmaceuticals, for the handsome sum of $30 million.

From his experience in developing medical and dental devices, Diamond formulated the following six design criteria:

- The KISS Principle: Keep It Simple, Stupid! The operation of the system must be simple and robust for usage by common people. Extensive training, tedious data interpretation, and multi-step preparation need to be avoided at all costs.

- Do not require the operator to calibrate the system. It must be self-calibrating. You want the operator and the physician to deal with the problem of the patient and not of your system.
- Never tell the doctor how to run his business. Present your new system as a significant improvement in the way that the doctor does things.
- Design for the minimum possible regulatory category (e.g., non-invasive versus invasive).
- If sterility is important in the clinical use of your device, it will lead to the opportunity to develop a sterile disposable sensor.
- Design for marketability and for distribution and service requirements.

4.6. Robotics and Artificial Intelligence

Many governments have made significant investment in the development of medical robotics (5). Other than the investment made to the development of industrial robots and in response to a shortage of 1 million caregivers by 2025, the Japanese government in 2013 granted $24.6 million to companies focusing on eldercare robotics. In the United States, President Obama launched the National Robotics Initiative in 2011 to stimulate development of robots for industrial automation, elder assistance and military applications. Operated by the National Science Foundation, the program has awarded more than $100 million contracts.

The private sector is also investing at increasingly higher levels (5). Google purchased Boston Dynamics, a robotics design company with Pentagon contracts, and DeepMind, a London-based artificial intelligence company for half of a billion. DeepMind is applying its expertise in artificial intelligence, machine learning and systems neuroscience to write the algorithm that operates a machine or a robot. One example is an algorithm that causes a computer to learn from mistakes in playing Atari 2600 video game, think in much the way that a human does in avoiding those mistakes, adjust actions through neural processes and, finally, develop a strategy to win the game.

In conjunction with the announcement in August 2017 that Foxconn would build a $10 billion factory in Wisconsin, the news also indicated that Foxconn had boasted that it had replaced 60,000 workers with robots at a single factory in China. The company even makes its own industrial robots, dubbed Foxbots, which work on its assembly lines. Foxconn was making about 10,000 Foxbots a year in 2015.

Japan leads the world in robotics, operating 310,000 of the 1.4 million industrial robots in the world. About 70% of total industrial robot sales take place in Japan, China, the U.S., South Korea and Germany. In 2013, 1,300 surgical robots were sold for an average price of $1.5 million each, accounting for 6% of professional service robots and 41% of the total sales value of industry robots.

Da Vinci and Raven Surgical System. This refers to the da Vinci surgical system that is manufactured by Intuitive Surgical in the United States. It was cleared by FDA in 2000 to perform minimally invasively complex surgeries, such as cardiac valve repair, and is used in more than 200,000 surgeries a year. The robot translates the movements of a surgeon's hand into more precise "micromovements" of the robot's tiny instruments. Smaller incisions leave the patients with less pain and shorter healing times. The arms of the robot swiveled 360 degrees, allowing it to do things that the human wrist cannot do, such as sew backward or sew underlying tissue.

However, at a cost of $1.8 million, it is available only to the wealthiest hospitals and institutions. Then there is the Raven, designed for the US Army and supported by NSF. It is a newer surgical robot that can test experimental procedures with open-source robotic operation codes. At $250,000, it is a much more accessible option than the Da Vinci systems. UCLA MAE Professor Jacob Rosen's RAVEN surgical robotic system was featured on the NBC medical drama "Heartbeat" during the episode "The Land of Normal," which aired in 2016.

With the expiration of some of Intuitive Surgical's patents, development of new surgical robotics already started heating up in 2016 and is likely to get hotter in years to come (3). For example, Stryker, Omni Life Sciences Inc., Zimmer Biomet, Medtronic, and Smith & Nephew are making progress with robots for total knee procedures.

Elder Care Robot. In early 2016, SoftBank released *Pepper*, a humanoid creature with the power to read and respond to human emotions. A massive breakthrough in our ability to connect with robots, Pepper could do wonders for the mental engagement and continual monitoring of those in need. The first crop of consumer-ready Pepper robots went on sale in Japan at $1,600 each. The entire supply of 1,000 robots sold out in less than a minute.

At present, in many medical and care facilities, two attendants or caregivers are required to move patients, or those in need of care, to and from their beds. There are data that show that nearly 70% of caregivers suffer from back pain due to the heavy burden on their back in lifting patients when assisting in normal daily activities.

In 2015, RT Works of Osaka developed the Encore Smart, an assisted walker that can take the elderly across rugged terrain. In 2016, the Riken Institute and Sumitomo Riko Company in Nagoya released their latest Robear, a robotic nurse in the shape of a bear that was gently lifting a patient of up to 176 pounds and carrying it around.

On why robots are the future of elder care? Mark Hay wrote on a website (4) "*As these technologies have developed in Japan, foreign companies have embraced them as well, with everyone from the European Union to Carnegie Mellon University to iRobot (the Roomba Company) sinking money into the development of robotic monitors, nurses, and companions for the elderly. In part, these firms are recognizing that so long as robots are accepted, they are necessary. But some are just recognizing how profitable the eldercare robotics market could be. Japan alone expects their market to grow from $140 million to $1 billion a year within a few decades. And Cyberdyne attracted a great deal of attention with its initial public offering last spring, raising $89 million to continue their work with exoskeletons.*"

4LUKE Arm. Developed by inventor Dean Kamen and DARPA, the LUKE arm offers a level of dexterity and flexibility that was almost unheard of in prosthesis technology. The LUKE (named for the robotic hand Luke Skywalker receives at the end of *The Empire Strikes Back*) features motors at the shoulder, elbow and hand. The LUKE Arm, the first device that intuitively moves multiple joints at one time enables wearers reach overhead, lift relatively heavy objects off the floor and handle delicate and more rugged objects with relative ease.

Electromyogram (EMG) electrodes pick up electrical signals from the patient's muscles, letting the wearer control the LUKE Arm's movements with muscle movements. The device transmits grip feedback to the user through a grip-force sensor. That thought-to-action control has led many to dub the LUKE Arm a mind-controlled device.

Cyberknife. This was introduced in 2001 as a robot-controlled radiosurgery system for the treatment of tumors. Instead of a scalpel, a bundled, high-energy X-ray beam is guided around the patient by a robot arm so that the dose is concentrated in the tumor. The patient with markers on his/her body lies freely on a robotic-controlled treatment table. A digital imaging system records the positions of the markers and communicates any movement – such as motion due to breathing — to the robotic system to compensate for these motions. In this way, the Cyberknife is able to radiate tumors with pinpoint accuracy, leaving the surrounding healthy tissue undamaged and not requiring hospitalization for treatment.

4.7. Medical Imaging and Information Technology

In a single year, a cancer research hospital generated more than half a million medical images in their fight against cancer. With the implementation of EHR system, the hospitals of U.S. will have billions of images available for processing. These images will be analyzed with the following goals (2):

Early Detection. This is a key indicator when treating cancer. Doctors can quickly compare and contrast images of a tumor with hundreds of thousands, or even millions, of other tumor images while diagnosing and treating cancer. With information technology such as big data analysis, Internet of Things (IoT), cloud computing and new deep learning techniques, a doctor can analyze the image of a tumor and set the course of treatment for quick patient recovery.

Improve Accuracy and Patient Outcomes. For example, the system can be a smart connected CT scanner that sends data directly to an IoT cloud for processing. With learning algorithms like neural networks in the IoT cloud, the system can interpret a medical image based on what it has learned from hundreds of thousands of others, and identify indicators of cancer that would otherwise remain undetected. The results are almost immediate, and are readily available to the doctor for use in diagnosing and treating patients.

Scan Follow-Up. Currently, there is no process in place for you to learn what happens to patients after you review their scans and make recommendations. Through information technology, you will be able to circle back with the referring physician to see what impact your contribution had. It is advantageous to know what happens.

Performance Improvement and Peer Feedback. Learning how referring physicians use your diagnoses to treat patients can help improve your work. If you have access to data concerning your reading accuracy, you can make positive changes.

All practitioners, even those who have several years of experience, can benefit from hearing what their peers think of their work, Siegel said. The use of information technology will allowing radiologists to provide feedback about the quality of their colleagues' studies, their use of the radiation dose, and how well they center patients for studies.

Radiation Dose. Limiting radiation doses has been an industry-wide concern for several years, but measuring and tracking it through information technology is relatively new. This type of tool can detect excessive levels of radiation quickly, prompting you to make changes to your protocols.

4.8. Innovations and Improvements

Revolutionary Products vs. Improvements. The technology that is required to build Da Vinci Surgical System are new and revolutionary. Extensive research and development went into constructing and controlling the robotics that does the surgical operation under the control of the surgeon. Carrying out this level of R&D within a small business tests the limits of entrepreneurs who must raise significant sums of venture capital and build up innovative technology long before the marketing phase. Consequently, the price of the Da Vinci System can be considerable.

In contrast, when a new device represents an improvement on existing technology, the likelihood of it being developed successfully will be higher, the company's investment in R&D will be lower and the manpower necessary to do the job will be less. This tradeoff in risk often must be made at the expense of a decrease in the total potential market share of the new product, and a risk of a shorter period of market exclusivity. The differences between these two categories of technology are highlighted in Fig. 4.3.

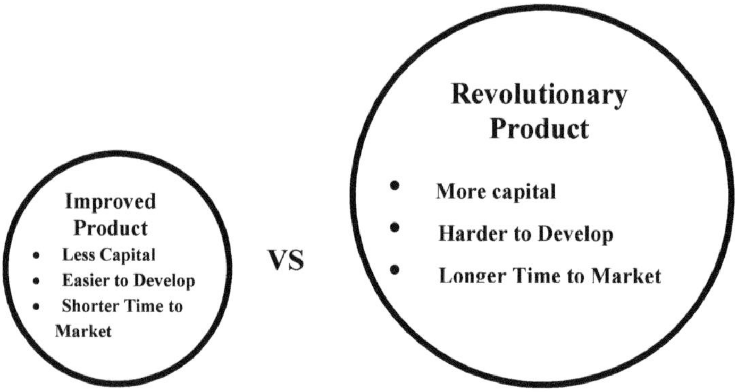

Fig. 4.3. Some key differences between improved and revolutionary products.

Between these two scenarios is disruptive technology, which may require less capital to develop than for a revolutionary product. A disruptive technology will involve higher risk and greater reward than a product that offers only a significant improvement.

Impact of Improvements on Growth. Webster's Dictionary defines "new" as "never existing before" and "innovation" as "something newly introduced." Although external pacemakers were already available in the 1940s, the drastically improved patient care due to the introduction of the implantable pacemaker by Greatbatch and Chardack certainly makes their pacemaker an innovation. As described in Section 1.4, Medtronic has made numerous improvements to the pacemaker for better performance and easier implantation. Medtronic's growth in pacemaker sales clearly demonstrates the importance of improvements to the financial success of today's Medtronic.

Improvement is essential for the continuous growth of a company. On the other hand, if you are a competitor, your financial success can be achieved more easily if you come up with an innovation to solve an important medical problem or an improvement that is revolutionary or disruptive.

References

1. Bakken E.E., *The Man's Full Life*, Medtronic Inc., pp. 45-54, 1999.

2. Chapa G, Carr C, Reynolds J, Practical IoT: Advanced Analytics for Medical Imaging, viewpoints.io/entry/practical-iot-advanced-analytics-for-medical-imaging, 2017.
3. Hartford J., 8 Medical Technologies That Will Be Huge in 2017, www.qmed.com/mpmn/gallery/image/medical-technologies-will-be-huge-2017
4. Hay M., Why Robots Are the Future of Elder Care? www.good.is/articles/robots-elder-care-pepper-exoskeletons-japan
5. Ross A., *The Industries of the Future*, Simon & Schuster, pp.15-43, 2016.
6. www.whcf.org/WHCF_WallaceHCoulter.htm

Chapter 5

Researching the Market for Your Invention

In this chapter, we first discuss the purpose of market research and then review websites that you can use to collect market information that is relevant to the medical problem that you intend your device to resolve. Consult also Sections 17.1 and 17.2, which help you to set the direction of your market research. This is necessary because of the special roles of your customers in the purchasing decision and the unique market characteristics of the medical device and pharmaceutical industry. Then we have the stories of two biotech ventures that were founded by Sheridan Snyder to illustrate his views on marketing and his thoughts and advice about entrepreneurship, building a company and market research. Finally we review the technology advances in cancer therapy for you to ponder on the market of cancer technology.

5.1. The Purpose of Market Research

Market research is necessary if you intend to obtain answers to the following three questions:

- What is the size and opportunity of the market that you are targeting?
- What differentiates your product from those of your competitors?
- What resources (time, personnel, and capital) will be required to complete the R&D and manufacturing phases and achieve your sales and profitability goals?

The second question is synonymous to competitive advantages over your competitors. Getting positive answer to these questions, which are highlighted in Fig. 5.1, will hopefully lead to positive market research results and the conclusion that it is worthwhile and feasible to develop your product.

For answers to the questions above you need:

- To interview people about your product's prospects,
- To check the literature on science, engineering and medicine for anything about your invention,
- To surf websites about disease and patient needs,
- To attend scientific, engineering, and medical meetings to identify your competitors, visit their exhibit booths and learn about development trends and any state-of-the-art research that is already underway.

Fig. 5.1. Positive market research results generated with positive information about market size, competitive advantages and resources that are required.

In addition, you should also be able to use similar search techniques to answer the following, related questions:

- On technology and competition:
  1. Will your product be functional and cost effective?
  2. Who are your current and future competitors?
  3. Will your product offer better performance than competitors' products?
  4. What is your invention's intellectual property position?
- On the clinical impact of your product:
  5. What is the disease or medical condition that your product addresses?

6. What is the potential size of the population of patients who could use your product?
7. What cannot be accomplished clinically without your product?

- On financing:
8. How long will it take to get it to market?
9. Are your competitors well-capitalized firms or small businesses?
10. What are the resource requirements for development?
11. Will you market your product globally or domestically?
12. How will you finance your operation in both the short-term and the long-term?

- On reimbursement and regulatory issues:
13. Are there any comparable products that are already approved and available in the market?
14. What regulatory approval processes do you anticipate that your product must undergo?
15. Once approved by the FDA, what reimbursement mechanisms will be utilized for your product?

By doing research to answer these questions during the Assessing Phase, you will also obtain ideas about the number of devices that you can sell and a price that would be competitive so that you will know whether your company can become profitable. With the feasibility study at the Assessing Phase done, you will also have an estimate of the manufacturing cost of your product. With more marketing research, you will obtain more data with which to prepare a financial projection for your company. Once you reach the Launching Phase, you may obtain additional marketing information that will enable you to develop an attractive business plan and then raise the funds required to manufacture and market your product.

5.2. Websites and Literatures on Medicine and Engineering

There are many nonprofit organizations that provide information about diseases or the body's organs. They include the following:

- American Diabetes Association (www.diabetes.org)
- American Heart Association (www.americanheart.org)
- American Liver Foundation (www.liverfoundation.org)

- National Kidney Foundation. (www.kidney.org)
- National Brain Tumor Foundation (www.braintumor.org).

The AMA (www.ama-assn.org) is another powerful resource. It lists on one website the names of 105 medical specialty societies, such as the American College of Cardiology, the American Academy of Orthopedic Surgeons, and the American College of Anesthesiology, all of which will provide more information.

In addition to medical specialty societies and the societies of medical and biological engineering (listed in Chapter 2), there are many professional societies that have a specific focus on the fields of medicine and engineering. One example is the North American Society of Pacing and Electrophysiology (NASPE), which hosts annual scientific sessions in various U.S. cities.

At NASPE's 24th annual scientific conference, which was held in May, 2003, in Washington D.C., 111 companies exhibited their pacing and electrophysiology products. Their names can be found on the society's website (www.naspe.org), along with a brief description of what they do. Guidant Inc., for example, exhibited at the 24th annual session and described itself as "a world leader in the design and development of cardiovascular medical products. Our devices help patients with heart disease return to active and productive lives."

Another exhibitor at the NASPE 24th annual scientific session was Medtronic Inc. The author was amazed to learn that Earl E. Bakken, who co-founded Medtronic Inc. in 1949 (and has served as its emeritus chairman since 1989), attends NASPE meetings regularly and was manning the Medtronic exhibit, where the author had an opportunity to interview him for this book.

In addition to being able to meet company representatives and size up the competition, entrepreneurs who attended the NASPE event had an opportunity to sit in on educational sessions involving cardiac resynchronization therapy, sudden cardiac death prevention trials, and patient selection and techniques for atrial ablation. Clearly, involvement with organizations like NASPE can provide the resourceful entrepreneur with a wealth of useful information.

Consider this additional example. The National Kidney Foundation website (www.kidney.org) provides the following statistical information:

- More than 20 million Americans — one in nine adults — have chronic kidney disease. More than 20 million others live at increased risk. Kidney disease is one of the costliest illnesses in the U.S. today.

- More than 400,000 Americans suffer from chronic kidney failure and need an artificial kidney machine to stay alive.
- Diabetes is the leading cause of chronic kidney failure. It accounts for approximately 44% of new cases of chronic kidney failure in the United States each year.
- Uncontrolled or poorly controlled high blood pressure is the second leading cause of chronic kidney failure in the United States. It accounts for about 35% of all cases.

With statistics like these, an entrepreneur can readily estimate the size of the patient population that would potentially benefit from the use of a new device to alleviate a certain kidney disease, or better diagnose the cause of kidney failure.

The following government-wide websites will be of interest to the biomedical entrepreneur:

- National Institutes of Health (www.nih.gov). The NIH is the steward of medical and behavioral research for the nation.
- Small Business Administration (www.sba.gov). One of the SBA's goals is to maintain and strengthen the nation's economy by aiding, counseling, assisting and protecting the interests of small businesses.
- U.S. Patent and Trademark Office (www.uspto.gov). The USPTO promotes industrial and technological progress in the United States and strengthens the national economy.
- Food and Drug Administration (www.fda.gov. The FDA promotes and protects the public's health by helping safe and effective products to reach the market in a timely way and monitoring products for continued safety after they are in use.
- Global Network of Environment & Technology (www.gnet.org). This site links interested parties to technology programs and agencies in all 50 States. Some of the listed agencies provide financial assistance to small businesses and start-up companies.

More information on how these three agencies — USPTO, Small Business Administration (SBA) and NIH — can help your venture development is provided in Chapters 7 to 10.

Following web links and doing Google searches will also yield useful information for your market study. Additional sources of market information are the research reports that Wall Street research analysts publish. Analysts frequently publish industry overview reports for targeted markets (i.e., breast cancer or implantable orthopedic devices). These reports typically include statistics that quantify the size of a market, giving assessments of competitors, and providing in-depth analyses of publicly traded companies within that market segment.

5.3. Two Biotech Ventures that Sheridan Snyder founded

Many people acknowledge Snyder as a legendary entrepreneurial figure in the biotechnology industry because of the enormous success that has been enjoyed by the biotechnology firms that he established. Sheridan Snyder graduated from the University of Virginia in 1958 with a degree in French and Linguistics. Two years later, his father died, and he learned that the family business had been far less successful than he had believed. Determined to turn the family's fortune around, Snyder launched his own career in business at the age of 24. Since then, he has founded 17 companies.

In this section, we look at two of his companies, Genzyme and Upstate Biotechnology. This will help you to learn how Snyder worked to resolve the special marketing opportunities that these two ventures offered.

Genzyme. Gaucher's disease is a very specialized disease that is prevalent among Jewish people of Eastern European ancestry. It is an enzyme-deficiency disease. Without the metabolizing enzyme called glucocerebrosidase, the body's organs become enlarged and, in severe cases, may stop functioning properly, causing death. Gaucher's disease kills thousands every year.

After many years of research, scientists at the National Institute of Health finally obtained a purified glucocerebrosidase enzyme that offered the potential to replace the deficient enzyme in people who have Gaucher's disease and restore them to good health. Unfortunately, no pharmaceutical company was interested in producing the purified enzyme as a therapy because the market for it was too small — as the disease affects fewer than 10,000 people worldwide.

NIH asked Snyder if he had an interest in developing the therapeutic enzyme for treating Type 1 Gaucher's disease,[a] the most prevalent form of the disease. In 1981, he co-founded Genzyme and ten years later the company received FDA approval for Ceredase® (alglucerase injection), the first effective therapy for Type 1 Gaucher's disease. The therapeutic enzyme represented a scientific breakthrough, because it was shown in clinical trials to not only relieve symptoms, but also to reverse the disease. In 1994, Genzyme received FDA approval for Cerezyme® (imiglucerase for injection), an improved therapy for Type 1 Gaucher's disease. It is now used to treat 95 percent of patients and is available in 60 countries.

While the development of the original therapy, Ceredase®, was good news to persons who were suffering from Gaucher's disease, it also presented Snyder's young company with a difficult question: How much money could Genzyme charge each patient for the drug?

Because of the high cost of developing the therapy and producing the enzyme, and the limited number of patients who needed it, Genzyme had to charge $140,000 per patient per year in order to make even a small profit. However, a price that high raised important ethical questions: What if a patient couldn't afford the therapy? Should he or she be denied it? At the time, no insurer was prepared to reimburse that amount of money to a patient who needed to use the therapy, although Genzyme had the proof that was necessary to justify its fee. Recognizing the benefit to patients, the U.S. Congress through Medicare and many non-profit organizations stepped in. As a result, the therapy for Gaucher's disease, despite its high price, was approved for reimbursement — although not without a lengthy battle first.

Genzyme has broadened its biotechnology to develop therapies for patients with many difficult, debilitating diseases. It is now a company of 6,000 employees in Boston and has annual revenues of $1.3 billion.

What is the lesson? You don't need to have a huge market in order to have a hugely successful idea. If your product is the only one that is capable of treating or curing a rare disease, you can still find a way to improve health and make a profit. As a case in point, the FDA has a specific provision entitled the Orphan Drug Act, which is designed to promote the development of drugs that target rare, serious conditions.

---

[a] Type 1 Gaucher's disease does not involve the central nervous system. Types II and III are rarer forms of the disease and symptoms often include problems of the central nervous system.

The Act is intended to promote the development of new drugs in clinical areas far outside of blockbuster market segments.

Upstate Biotechnology. Snyder became involved with Upstate Inc. in 1998 as a result of the company's attempt to sell out to a non-profit foundation. After a careful market analysis, Snyder decided to merge his company, Argonex, into Upstate and served as the Chairman and CEO of the new company.

In subsequent years, Upstate has obtained licenses from 220 institutions to manufacture novel, leading-edge proteins and reagents that are used to gain a better understanding of protein-to-protein interactions, the function of proteins within the development of disease, and the role of protein in cell signaling. Discoveries in cell signaling and advances in drug discovery may revolutionize medicine in the coming years. The company employs about 220 people and is headquartered in Charlottesville.

Upstate has developed a method to attach different proteins to one microscopic sphere (appropriately known as a "microsphere") and then screen the various reactions of the proteins with laser-based technology. This method has transformed the way in which people are tested for allergies. When you tested someone for allergies in the past, you needed a different test for each allergen. Now, with this Upstate technology, you can put all of those screening proteins onto one microsphere, and test for all of those allergens in one pass with one drop of blood. In essence, Upstate's new test method will be used by everyone who has allergies and at a much lower price than afforded by other systems. As 55% of people in the US have at least one or two allergies, it is not difficult to estimate the probable market size for such a screening device.

Before too long, there will be biochips with test sites in microarrays that enable many tests to be performed at the same time. They will have more screening power than is possible with microspheres. Twenty-five years from now, Snyder predicts, you'll be able to go down to the drugstore and purchase a $10 bag of vitamins, each of which will contain its own biochip to permit the screening of large numbers of biological and chemical compounds for disease diagnosis to the detection of bioterrorism agents. Microarray development is certainly within the domain of biomedical engineering. With such a high-potential idea, it is an exciting time to be in biomedical engineering.

Snyder believes that his primary responsibility to shareholders is to create value for them. This is the only criterion by which he evaluates himself as a company's CEO. The fact that Upstate, which was valued at $5 million in 1998, was sold in 2004 for $205 million to Serologicals Corporation means that he is doing his job very well.

The sale of Upstate resulted in a gift of $45 million to the University of Virginia Health Center to build a new children's hospital, the Emily Couric Cancer Center and the Sheridan G. Snyder Translational Research Center.

5.4. Entrepreneurial Advices and Thoughts from Snyder

According to Sheridan Snyder, entrepreneurship is simply the translation of science and engineering to commercial application. In the lecture that he delivered to the class in BME entrepreneurship at the University of Virginia in February, 2003, he addressed many interesting questions: How does a venture develop? What forms the basis of it? Who are the parties involved? What is venture capital? How does someone become involved in a venture? Should you think of yourself as an entrepreneur? Some highlights from his talk are given below.

Snyder admitted that, despite his advancing age, he couldn't get over the habit of building biotechnology companies. He enjoyed developing and expanding new things. It was especially enjoyable to be able to "*get up in the morning, knowing that you will be developing a new cancer therapy using a kinase to block the mutation of a certain breast cancer gene*," he said.

He encouraged students to follow his lead in creating biotech companies. "*There's no law that says you can't be an entrepreneur,*" Snyder said. "*There's nothing that says you can't raise money. There's nothing that says you can't create a business plan. The only thing that will stop you is a straight line between your two ears. The only thing you need is the conviction that you, indeed, can do it.*"

Snyder recalled a time when a bright MBA student from the University of Virginia asked whether he stood a better chance of becoming a successful entrepreneur if he first went to work for the management consulting firm of McKinsey & Company or another consulting firm. Snyder's advice to him was "*Go to either company.*" Once a would-be entrepreneur learns everything possible about running a business from working for a good company, "*the entrepreneur, by instinct, will go straight from there to file incorporation papers and set up his or her business,*" he said. "*You'll find your way, you'll survive, and you'll succeed.*"

According to Snyder, there are only two reasons to fail in developing a company. "*One of them is if you die,*" he said. "*The only other reason is incarceration. If they put you away for 45 years, you're not going to be able to develop your company.*" Except for these two reasons, there is no reason for failure. "*Don't say, I can't get money, or I can't do this or that.*"

Here's another tip that Snyder gave to the class. If you want to start a biotech company, then building up a solid reputation as a successful entrepreneur is important. As an example, he cited the friendship that he has developed through the years with Sydney Brenner, a molecular researcher who received the 2002 Nobel Prize in Physiology or Medicine for his discovery of the RNA messenger, and who has worked with James D. Watson and Francis H. C. Crick, the two scientists who discovered the structure of DNA at Cambridge University. The brilliant Brenner was the first person to recommend Snyder to a venture capitalist when Snyder was starting up Genzyme. It was "the funniest thing," Snyder said, "*to hear a future Nobel Prize winner telling the Rothschild Fund to invest in Sherry Snyder* Snyder, Sheridan — *someone without an advanced degree, any science background, and who had never invented anything in biotech in his life — simply because Brenner thought that I would succeed in developing a biotech company.*"

Snyder expressed his hope that the students would one day latch on to some interesting biomedical engineering idea, such as biochips that float in your bloodstream and emit therapies. "*Just let your imagination go,*" he said. "*The world already has all sorts of information that you can use to develop your idea for biomedical engineering entrepreneurship. Think of biomedical imaging. Some of the images are cell-based and some are molecular-based. They are just incredible. Soon you'll be able to connect biomedical imaging with therapies.*"[b]

However, imagination will carry you only so far. "*It's important to understand your market before you go too far down the development track,*" he added. "*For those working in the world of science, you're emotionally excited about the science, and you are attached to the*

---

[b] The magnetic guidance system for less invasive surgery that is described in Chapter 3 is such an example.

science. You feel you are going to re-do the world. That makes you a little bit blind. Sooner or later, the market is the judge of whether or not your product is going to succeed."

"*Why wait for the judge?*" he asked. "*Understand the judge at the front end, and your chances for success are going to be a lot better. The judges for a company are the investors. They're going to vote for what you're doing, especially if you have done a thorough job of the market study.*"

As a venture capitalist, Snyder wants to see a market study that reflects a minimum of 50 interviews with relevant experts — half conducted by telephone and half in person. Such extensive interviewing helps to ensure that accurate assumptions will be made about the competitive environment for a particular technology. "*No matter what you think your science is, it's going to have to move out into the world and compete against others,*" he said. "*The sooner you know what the competition is, the better off your company will be.*"

Another thing that creates value for a new company in the eyes of investors is attaining milestones. "*Science is nebulous and fuzzy during its development period, especially in academic research,*" Snyder said. "*Even though you think you have achieved a great advance in science, its value may stay within the university. Don't ever think that you can make a business merely by selling or licensing technology. A lot of people have tried to do that and have actually failed. If you can't put your product in a box and send it out with an invoice and sell it, you haven't got a business.*"

Snyder reminded students that, in biotechnology, the market pulls and the science pushes. "*This is something you've heard many times,*" he said. "*It is a big temptation when you're involved in science, when you're involved in biomedical engineering, or when you're involved in protein chemistry*" to get involved in the science push without considering the market pull.

Two weeks prior to delivering his lecture to students at the University of Virginia, Snyder had been at University College in London, where the leading researchers there talked with him about the regulatory role of nitric oxide. While he admits that getting his mind around the science can sometimes be challenging for a former French major like himself, he has no trouble in understanding the market for cell-signaling proteins. Because he knew that a market exists, he was willing to make a 10-year agreement with University College in London to produce the cell-signaling proteins that had been invented there.

"*The worst time to figure out your market is when your time for research and development has run out,*" Snyder says. "*If you understand what market you're aiming for from the beginning, then your science is going to push and the market will pull you to the right direction.*"

What makes a company succeed? It requires more than a good idea or a scientific discovery. "*You've got to have a product, and you've got to have enough science behind the product that you can leverage it to build the company,*" he said. "*You have to ask yourself this: Do you have a product broad enough to create a whole company? Don't get sucked into building a company when really what you have is a relatively narrow piece of science.*"

A strong intellectual property will help improve a company's chances for success. "*It's very difficult in the early stage to try to develop a company without an intellectual property or a patent,*" Snyder said. "*Patenting your invention can be done with a small investment. But it's not easy.*"

"This *is why entrepreneurship in biomedical engineering calls for partnership between business people and engineers. They've got to work together to understand the market, not merely the science,*" he said.

Snyder likes to work with small companies, which often have the feel of a family as employees grow close to one another. "*The whole team is trying to accomplish something,*" he said. "*It's the same type of pioneering that mountain climbers get by going out together to the peak of Mount Everest.*"

After starting up 17 companies, Snyder still has a hard time to let them go, precisely because he becomes attached to each company and the people who work for it. Nevertheless, if you've done your job well, he believes, eventually there will be a time to exit. Usually that time comes when the company's value has become so high that there is a risk of someone attempting a takeover. "*You go public, you sell, you merge,*" he said of the exit strategies available to the entrepreneur.

BioCatalyst International is Mr. Snyder's the most recent business enterprise. In partnership with leading scientists, BioCatalyst forms a series of start-ups to move the scientists' ideas beyond the boundaries of science into profitable biotech companies.

"*If you keep looking around, you'll start to find that all of a sudden, something will make sense to you,*" said Snyder, who added that most of the ideas for his ventures have come from his associations with venture capitalists and renowned scientists.

"*If you like to take risk, like the chance to succeed, and don't want to be bored, a career in building companies will be the right place for you, although it will scare the pants off you at one time or another*," he concluded.

5.5. Technological Advances in Cancer Therapy

The goal of radiotherapy is to deliver as much dose as possible to the tumor while sparing normal tissue from damage. Technological advances incorporating new imaging modalities, more powerful computers and software and new delivery systems, such as advanced linear accelerators, have helped to achieve this goal. Three advances are described here.

Focused Ultrasound Treatment. This technology was first conceptualized in 1950. Due to the advancement in magnetic resonance or ultrasound imaging, human treatments using focused ultrasound have exploded in the past decade. A concave transducer or an array of transducers can focus their acoustic power onto a small, "cigar"-shaped volume of tissue, typically a few cubic millimeters, at an intensity of 100 to $10,000\ \mathrm{W/cm^2}$ and a peak compression pressure of up to 30 MPa. In comparison, diagnostic ultrasound transducers deliver intensities of $0.0001–0.1000\ \mathrm{W/cm^2}$ and a compression of 0.001–0.003 MPa. Rapid elevation of the local tissue's temperature is the main causative mechanism of tissue destruction.

Focused ultrasound treatments can be administered on an outpatient basis, require no incisions, and may cause minimal discomfort and few complications and allowing for rapid recovery. Nearly 168,000 patients have been treated with focused ultrasound. Most (i.e., 80%) of these treatments have been for prostate diseases and uterine fibroids. For prostate and pancreatic cancer, randomized controlled trials are being done to compare focused ultrasound treatments to standard radiation and chemotherapy treatments. The U.S. has 78 hospitals doing these clinical trials.

According to the Focus Ultrasound Foundation, there are 45 manufacturers of focused ultrasound devices worldwide. They are based in the U.S. (12), China (9). France (7), Taiwan (3), Canada (3) and other countries (11).

MasSpec Pen. This pen is a hand-held device that enables doctors to determine in real-time whether tissue is surgically clean of all cancer cells surrounding a tumor. It delivers results in about 10 seconds, according to researchers of the University of Texas at Austin.

In particular, cancers such as those of the breast, pancreas and brain "tend to invade surrounding normal tissue," the report's author Eberlin said. "For a surgeon, it can be really tricky, because these cells resemble normal tissue. Just judging by the eye, it can be very challenging."

The single-use pen connects by a thin tube to a mass spectrometer, a scientific device that analyzes the chemical and mass composition of molecules. The pen's accuracy currently stands at more than 96%, based on tests that used tissues removed from 253 cancer patients.

Cancer surgeons these days send tissue samples off to an in-hospital pathology lab during the operation to see if they have removed the entire tumor. However, the process can take as long as a half hour, during which time the patient remains on the table and under anesthesia. Even with this precaution, most surgeons have had the misfortune of discovering that tissues that were initially judged to be cancer-free by pathologists during the operation were later ruled to be positive for cancer during a follow-up analysis.

Radiotherapy. To show the progress, three new modularities of radiotherapies are described below:

- ***3D Conformal Radiotherapy (3DCRT).*** 2D radiation therapy using rectangular fields and plain X-ray imaging has largely been replaced by 3D radiation therapy based on CT imaging that permits accurate localization of the tumor for optimal beam placement and shielding.
- ***Intensity Modulated Radiation Therapy (IMRT).*** IMRT enables the oncologist to create irregular-shaped radiation doses that conform to the tumor while simultaneously avoiding critical organs. IMRT is made possible by: a) inverse planning software and b) computer-controlled intensity-modulation of multiple radiation beams during treatment. This has permitted improvements in the therapeutic ratio (the ratio of the maximally tolerated radiation dose to the minimal curative or effective radiation dose).
- ***Image-guided Radiotherapy (IGRT).*** As treatment margins become tighter and more conformal, the potential to miss a

tumor due to organ motion and patient setup variations become greater. When critical structures are close to the tumor, a slight positional error may also lead to inadvertent radiation of the normal organs. IGRT permits the detection of such errors by information acquired through pre-radiotherapy imaging, which allows for correction.

Chapter 6

Forming the Company

6.1. Rationale for the Corporation

When tax and liability are taken into account, you may decide to establish your company as an S Corporation. Before receiving any venture capital or going public, you can change it to a C Corporation.

Although the cost to establish a sole proprietorship, partnership or limited partnership is relatively low and the operation of each is simple, they are bad entities for most business operations as they do not offer any liability protection for the owners. This means that your business, house, and personal property can be lost in legal liability suits. A Limited Liability Company (LLC) does not have a problem of liability. As an LLC does not issue stock, the IRS treats an LLC as a partnership. If your LLC is registered in Nevada and you are a resident of California, you must be cautious because you might be liable for payment of the same onerous fees as you would for a California LLC. The LLC is a relatively new entity. Many legal issues concerning LLCs have not been fully tested in the courts yet. Some venture capitalists do not like to invest in LLCs.

An S Corporation operates under flow-through taxation, meaning that any gain or loss is passed on to the corporation's shareholders. An S Corporation has no income tax to pay. Within certain limits, a company loss flows through as a loss of the shareholders when they calculate their taxable incomes. In contrast to a C Corporation, an S Corporation is subject to the following four restrictions:

- The shareholders can be only individuals, estates and certain trusts,
- The total number of shareholders of an S Corporation may not exceed 75,
- The S corporation may have only one class of shares. Differences in voting rights are allowed,
- No shareholder may be a non-resident alien.

If your S Corporation violates any of these restrictions, the IRS will notify the termination of your S Corporation status and convert it to a C Corporation.

A C Corporation is required to pay income tax. Its losses can no longer flow through to its shareholders. The C Corporation can use any after-tax gain for future growth and expansion. It can also distribute the gain as a dividend to shareholders. The dividend distribution to shareholders must be included in their taxable incomes for tax purposes.

To fully understand the consequence of this so called double taxation (i.e., the corporation pays tax once on its profit and the shareholders pay individual tax on the dividends they receive from the C corporation), we examine the case of a corporation that has a taxable income of $50,000. The corporation tax rate on this income or profit is 15%. When the shareholders received their dividends, they will pay individual income tax at the rate of 28%. For C Corporation, one scenario considered is the case (shown in the second column of Table 6.1) in which no dividends are distributed to the shareholders. In the other scenario (the third column), all after-tax profit is distributed to the shareholders as dividends. The fourth column describes the case of an S corporation. In the last two cases, the individuals reinvested their after-tax cash in the company. The final amount that the company has, because there was no dividend distribution or because the dividends were reinvested in the company, is termed the "amount remaining" in the company to enhance the capital that the company is holding for growth, acquisition, loan pay-back, and investor buy-out. The corporate tax, the dividend distribution to shareholders and the tax on individuals are listed in sequence as the second to the fourth row of Table 6.1.

Table 6.1. Amount remaining in a C or S Corporation with a profit of $50,000.

	C Corp. with no dividend distribution	**C Corp. with full dividend distribution**	**S Corp.**
Corporate profit	$50,000	$50,000	$50,000
Corporate tax	$7,500	$7,500	0
Distribution to shareholders	0	$42,500	$50,000
Individual tax*	--	$11,900	$14,000
Amount remaining	$42,500	$30,600**	$36,000**
Difference (%)		-28%	-15%

*At an individual tax rate of 28%.
**Reinvested in the corporation.

As seen in the last row of Table 6.1, the first scenario of a C Corporation with no dividends declared or distributed has retained the greatest amount of amount remaining. The amount remaining is 28% lower for the C Corporation that distributes all of its profit as dividends to the shareholders and then receives the after-tax amount back as reinvestment in the company. The S corporation that received reinvestment of all profits had an amount (of profit) remaining that was 15% lower than that of the C Corporation that declared and distributed no dividends.

In Table 6.2, we show the impact of double taxation for a company that has a profit or taxable income of $10 million. According to the tax rate schedule for corporations, the tax is $3.4 million. For the dividends distributed to the shareholders, we use an individual tax rate of 35% for the calculations. As seen in the last row of the table, the amount remaining for the C Corporation that distributes all of its profit as dividends is 67% lower than the case of the C Corporation that had no dividend distribution. On the other hand, the amount remaining in the S Corporation is only a negligible percentage lower than the case of C Corporation that distributed no dividends.

Table 6.2 Amount remaining in a C or S Corporation from a pre-tax profit of $10 million.

	C Corp. with no dividend distribution	C Corp. with profit distributed as dividends	S Corp.
Corporate profit	$10,000,000	$10,000,000	$10,000,000
Corporate tax	$3,400,000	$3,400,000	0
Distribution to shareholders	0	$6,600,000	$10,000,000
Individual tax*	--	$1,190,000	$3,500,000
Amount remaining	$6,600,000	$2,210,000	$6,500,000
Difference in %		-67%	-2%

*At an individual tax rate of 35%.

After any investment in capital assets (buildings and, machinery) and acquisitions, the cash remaining in the company is termed accumulated earnings or retained earnings. Tax complications may arise if your C Corporation has too much accumulated earnings. Although your company may claim that the accumulated earning is kept as working capital or as a reserve for an economic downturn, it may be up to the IRS, and then the court, to decide whether your company will be taxed an additional 40% in excess of the company's regular tax rate.

The reason for such a penalty is that the IRS wants your C Corporation to distribute dividends so that IRS will have an opportunity to tax the “profits” again. Although the tax courts have the final say on such an issue, it is much better for you to consult your CPA and attorney on the subject of how you distribute dividends and handle your accumulated earnings in order to protect your company.

If the company’s operating capital comes from individual angel investors, your company can remain as an S Corporation. If you need a capital infusion from venture capital firms and/or other companies in exchange for shares in the company, these firms will become owners of the company and the restriction on shareholders will require your company to become a C Corporation.

If any large company owns more than 50% of your company’s shares, your company (even if it has fewer than 500 employees) may no longer qualify to receive government assistance as a small business.

The owners of an S or C Corporation are the shareholders. Some shares have voting right and others do not. Only C Corporations can have more than one class of stocks, such as common shares and preferred shares. Preferred shares often have priority over common shares in the payment of dividends or the distribution of assets on liquidation of the company.

The corporation files an article of incorporation with the state and has by-laws that govern its operation. The by-laws stipulate, for example, the choice of bank and auditor that the company will use, who have signing authority (the authority to make financial commitments on behalf of the company), who are the Directors of the Board and who has been elected to the positions of Chairman of the Board, Chief Executive Officer and President.

In summary, the filing of a corporation may be a simple technicality, but operating the company for optimal taxation is complex. In this section, we cover only a few operational features of a C or S Corporation. To set up a corporation that is right for you, to manage it in accordance with federal and state laws and to protect and advance your business interests, it is important that you work with advisors who are knowledgeable in legal, tax and accounting matters of a corporation.

6.2. Organizational Structure

You may start your company as a two-person company with a CEO and a chief technology Officer (CTO). The CEO is responsible for the business development and the CTO is responsible for product R&D. They work together to map out the vision, strategy and plan to grow the company. Depending on the situation, the CEO and CTO may be the same individual.

As the company moves into the Assessing Phase, an entrepreneurial team is established to carry out the items described in Chapters 3 to 5 and also the feasibility study, which will demonstrate that the device can be built with current technology. The management may change to reflect the state of the company at the Assessing Phase.

As the company proceeds to the Launching Phase, you hire more people to do the jobs as described in Chapters 6 to 11. You may invite those entrepreneurs, experts and physicians whom you interviewed previously to serve on Advisory Board. Because this Board is advisory, the liability of its directors is much less than the directors of a regular Board of Directors, which has the sole authority to hire (and fire) the CEO of the company and oversee all activities of the company. As more funds are required to operate the company, it may be desirable to hire a CFO to manage the company's finances. The R&D activities also need to expand to include work on patent and regulatory affairs. You may consider having a semi-formal organization as depicted in Fig. 6.1.

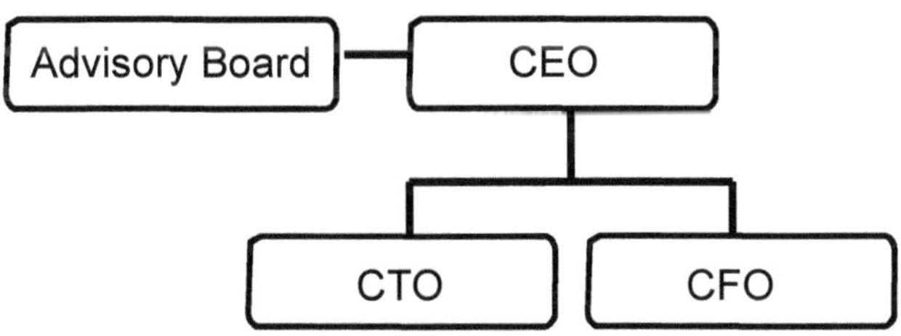

Fig. 6.1. Organization chart of a company in its Launching Phase.

When the company reaches the Building-up Phase, the company will add a VP of Manufacturing with responsibility also for quality control and regulatory affairs and a VP of Marketing and Sales to promote the product and manage the company's sale activities. A Board of Directors that reflects the interest of stockholders and investors in the Company should be established. The company's organization may have a structure like that in Fig. 6.2.

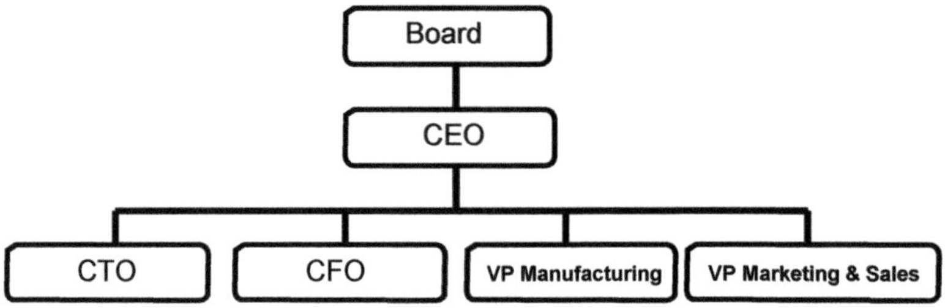

Fig. 6.2. Organization chart of a company in its Building-up phase.

6.3. Capital required for the company's operation

Phases in Development. As described in Chapter 10, your company can apply for SBIR grants to fund its operation. NIH will fund your Phase I feasibility study with up to $100,000 for six months and the Phase II research and development work with up to $750,000 for two years. In this book, Phase I corresponds closely to the Assessing Phase and Phase II corresponds to the Launching Phase.

In view of the NIH funding levels for Phase I and II, we suggest that you may need to raise $50,000 to $150,000 to accomplish the tasks in the Assessing Phase and $750,000 to $1,500,000 for the Launching Phase. If you can raise this much money for the company's operation, it is likely that your entrepreneur team in the Assessing Phase will have two to four team members at most. The officers and staff of the company may grow to five to ten members during the Launching Phase. In the Building-up Phase, the complexity of manufacturing and marketing the product may require that you raise $2 to $4 million and hire 15 to 20 staff members. These requirements are summarized in Table 6.3.

Table 6.3. The time required, the investment sought and the employees hired during three phases of company development. The values listed are only suggested and you can expect large variations, depending on the complexity of the development.

	Assessing Phase	Launching Phase	Building-up Phase
Time Required	6 months to 1 year	2 to 3 years	3 to 5 years
Investment Sought	$0.05M to $0.15M	$0.75M to $1.5M	$2M to $4M
Employees Hired	2~4	5~10	15~20

Raising money is not only challenging, but also time consuming. A CEO must be careful in managing the money, find other revenues, and employ various means to reduce expenditures. For example, Medtronic, which Bakken founded, had income from its repair service and selling other companies' products. If a job can be done by a part-time employee or for a short period of time, you can consider outsourcing the job to a competent contractor, instead of someone in-house full time.

If your product is revolutionary, the complexity of R&D, the time, investment and employees required can be as high as five to ten times what appears in Table 6.3. For pharmaceutical products that require clinical trials, the time, investment and employees necessary for their development may be fifty to one hundred times more than those in Table 6.3. In any case, the development of a good business plan may help you to generate a more realistic estimate of the time, money and people needed for the Assessing, Launching and Building-up Phase.

Sources of Funding. The first source of funding may be the entrepreneurial team that founded the company. These persons can use their own savings, borrow from the family, secure loans from banks or obtain cash from credit cards. Bank loans will require collateral whereas credit cards do not. However, the interest rate of credit cards is much higher than that of bank loans. The funds that are obtained by the founders of the company will be termed "the founder's investment."

Second sources of funds are the family, friends or angel investors who want to invest in your company. We assume that all of these investors qualify as individuals for tax purpose and that the invested money is used to acquire stock at market value (the nominal value that was agreed upon as the market value at the time of investment). The funds so obtained will be termed "angel investment." Venture capital firms are the third source of funds. Shares are acquired by the venture capital firms at the same or different market value or price that was offered to angel investors.

When the company achieves a certain status that is defined by the stock market, your company can be listed on a stock exchange and traded in that market. The first time your company's shares are sold in a securities market is termed your company's initial public offering (IPO). Your company can also sell bonds in a bond market to raise operating cash. These two methods of raising money are not available to start-ups, because they are unlikely to meet the listing requirements of the stocks or bond market.

Chapter 13 provides a more detailed explanation of how to obtaining financing from angel investors and VC's, what they are looking for, and the due diligence they undertake to evaluate your funding request.

6.4. Company Registration, Share Distribution and Exit Strategy

Registering the Company. To get the company registered, the first step is to select the company's name. Make sure that the name does not infringe on the trademark of another company and has not been used as a corporation name in the state in which you are going to register and/or work. Next, obtain an employer-identification number from the Internal Revenue Service (IRS). For a small company that is owned by a few founders, the founders can serve as the various officers (e.g., president, secretary, treasurer) of the company. If it is to your benefit to have the income and loss of the corporation flow through to the owners (i.e., the shareholders), you will set up your company as an S corporation.

Discuss your plan to form the corporation with a competent attorney before registering your company in the state where you plan to operate. As an alternative, you may register your company in Delaware or Nevada in which there is no state corporation tax for out-of-state companies. After registration, you will receive a corporation certificate from that state. Your attorney will provide you a corporate seal, stock certificates, the by-laws of your company, and sample forms for recording the minutes of directors' meetings. The by-laws describe the way in which your company does business and how to elect and remove officers and directors of the corporation. Finally, you must designate an agent and register him or her with the state as this agent's service is an important aspect of protecting your corporation's status.

Company Location. Selecting the location where you will establish and build up the company is an important decision. First, it determines the challenge that you will encounter in recruiting the right workers. If your company is a high-tech or biotech company, San Diego, Boston and San Francisco are places to consider. This is because many similar companies have been established in these metropolitan areas. Community support, cost of living, local tax incentives, and the environment are other factors that you need to consider as you decide where to locate your company.

Investment and Stock Distribution. We use Company A in some hypothetical settings to illustrate the relationship between investment and stock distribution. Assume that the company's Board of Directors authorizes the issuing of five million shares and that the founders will invest $100,000 in the company. Soon after the Assessing Phase begins, the team convinces several angel investors to invest $300,000. With these two investments, the team is also able to win a Phase I grant of $100,000 for the Assessing Phase and later a Phase II grant of $750,000 during the Launching Phase. Suppose that the progress in R&D and the NIH funding convince a few venture firms to invest $2,000,000 to move the company to the Building-up and Commercialization Phase. This venture investment will normally be made in several installments as certain milestones are reached during the Building-up Phase. Over this course, the company's employees also invested $40,000. In total, the company will have funding of $3,290,000 with which to take the company to the stage of marketing its product for a profit.

In recognition of the work performed by the founders, the market value at which they will acquire their shares in the company will be set at $0.10. Thus the founders will receive one million shares for their investment of $100,000. After the company demonstrates the invention's potential, the market value of the share may rises to $1.00. Thus, the angel investors will receive 300,000 shares for their investment of $300,000. Following this angel investment, the company can remain as an S corporation if the angel investors are individuals and the total number of shareholders is less than 75. Equally important for the entrepreneur team, the company will remain in their hands because they own the majority of the company shares.

With more work accomplished by the company as it moves to the Building-up Phase, the market value may rise to $2.00 per share. The investment of two million dollars by the venture capital firms will enable the angel investors to purchase one million shares. In order to obtain the VC funding, you may have an agreement that states that the VCs have preferred shares and majority control of the Board of Directors. (What should be agreed upon in the agreement will be reviewed in Chapter 14.) At this time, the company needs to be converted to a C Corporation, because some shareholders are now corporations. As the company advances through the three phases, employees will be making a substantial contribution to the company's development and deserve to be rewarded with stock.

The market value/share assigned and the number of shares allocated for the founders, angel investors, venture capital firms and employees are listed in the third and fourth columns of Table 6.4. With two and a half million Shares already allocated, the company will have another two and a half million shares for an IPO or later rounds of venture financing.

If a second round of venture funding becomes necessary, the board may authorize the issuance of additional shares so that that the company can raise additional operating funds. In so doing, the shares that the founders and early investors own will be diluted as they will own a proportionately smaller share of the ownership of the company.

Table 6.4. Distribution of Company A shares to founders, angel investors, and venture capital firms.

	Cash received	Market value/share	Shares
Founders	$100,000	$0.1	1,000,000
Angel Investors	$300,000	$1.00	300,000
Venture Firms	$2,000,000	$2.00	1,000,000
Option Pool*	$40,000	$0.2	200,000
SBIR Grants	$850,000		0
Total	$3,290,000		2,500,000

* An additional 2.5 million shares are reserved for IPO.

Demands and Traps in Financial Transactions. The actual closing of financial transactions between the company and investors and the demands from angel investors and venture capital firms are much more elaborate and extensive than the simple transaction of stocks that is illustrated here. Demands from investors can include the appointment of certain directors to the Board and creation of many classes of stocks other than common and preferred stocks. Some demands can even be a trap in disguise that would enable investors to take over your company without due compensation. Chapter 14 explains what those demands may be and how to negotiate terms that are mutually acceptable and beneficial.

Valuation and Control. The Market value per share is the value of the company divided by the outstanding shares at that time. How does one put a "value" on a proposed business or a start-up? One way is to determine by accounting the value of tangible properties, such as land, buildings, equipment and patents that the company owns.

Another conventional approach is to set a price-to-earnings ratio and then use the earnings of the last fiscal year to calculate the value of the company. However, it is difficult to determine the value of the company before the company has had any earnings or only has the potential to make money in the future. The use of the price-to-earnings ratio applies mainly to well-established companies. Consequently, assigning a value to a company that has just been formed becomes a matter of sheer salesmanship by the entrepreneurs and the credibility of the business plan. Just like buying a house, the value of the company or the market value per share is whatever investors are willing to pay and the company owners are willing to accept.

Who holds control of Company A in Table 6.4? Although the venture firms have only about 43% of the shares, it is likely that their shares are preferred shares. However, they may have been able to succeed in negotiating an arrangement by which they will have four seats on the Board of Directors (i.e., have been able to select four of the Directors), with two seats for the founders and angel investors, and one independent seat (Director). In effect, they have legal control of the company.

In high tech businesses, the founders often bring to the table expertise that is essential to the operation of the business. Thus, the founders exert "control" even though they may own only a small fraction of the company and occupy at most two seats on the Board of Directors.

Exit Strategy. As previously agreed by the founders and the investors, angel investors and venture capital firms may at some future time wish to recover their investment and take their profit. The first thing to do is to ask a professional or CPA firm to estimate the value of the company based on its stock market performance. Suppose the value of the company is estimated to be a total of $25 million. This valuation implies that the 2.5 million shares, which the founders, investors and employees own, are worth $10 per share.

One way by which the company can repay the investors is by using some of its cash reserves and obtaining a bank loan using the company as collateral. If the investors and the company agree that the share price is

$10 per share, the value of the investments of the four categories of shareholders that appear in Table 6.4, are those that appear below in Table 6.5. The angel investors will receive three million dollars. The venture capital firms will receive ten million dollars. After these two groups "cash out," the founders and employees retain the company as an employee-owned, private company. In this exercise, the founders and employees receive the greatest gains because of their extraordinary effort in getting the company to a point at which its share price is $10.

Table 6.5. Increases in value of holdings at a price of $10/share in Company A.

	Value of Initial Investment	Shares Owned	Final Value of Investment
Founders	$10,000	1,000,000	$10,000,000
Angel Investors	$300,000	300,000	$3,000,000
Venture Firms	$2,000,000	1,000,000	$10,000,000
Employees' Options	$40,000	200,000	$2,000,000

Suppose the angel investors made their investment seven years ago and the VC invested five years ago. Then, the annual rate of return would be 39% for angel investors and 38% for VC.

If you have a good valuation for your company, it is time to pursue an IPO to obtain more cash for the company with which to repay the angel investors and venture capital firms. Then, your company will become a public company. Of course, there are many regulations for a public company to comply with.

As the third exit strategy, hopefully you can sell the company for $25 million. In that case, the venture capital firms will recover their investment and a 500% return, and the angel investors will recoup their investment and a 1,000% return. Please note that the hypothetical amounts used for the calculation are the net proceeds of a bank loan, IPO or sale of the company. Make sure that you have a team of qualified advisors to carry out ethically and legally the stock distribution and the exit arrangement.

Chapter 7

Patenting your Invention

7.1. The US Patent and Trademark Office

The basic role of the US Patent and Trademark Office (USPTO) is to promote the progress of science and useful arts by granting patents to protect inventions over limited periods of time. Primarily, patents prevent public disclosure of your invention from reducing your opportunity to achieve economic gain from its development and sale. Secondly, patents inherently encourage public disclosure by safeguarding the inventor's rights to economic gain. Thus, they support the free exchange of new ideas, so that other inventors can build on the patented device or process to promote further technological advances. Congress believes that the patent disclosure process is very important for expansion of the nation's economy.

A booklet that explains patents and the USPTO's operation is available at its website at www.USPTO.gov. In this chapter, we review the basics of patenting and examine a number of patent issues of interest to entrepreneurs.

7.2. Utility, Design, and Plant Patents

There are three types of patents. They are: utility, design, and plant.

Utility Patents. A utility patent is granted to someone who invents a new and useful process, machine, article of manufacture or composition of matter or any new useful improvement. Since a new medical monitoring device is a machine, its patent will fall into this category.

When you invent a new process to measure a medical parameter, such as blood pressure, hematocrit or glucose concentration in blood, you can elect to register the process as a utility patent. This type of patent also covers new drugs, new man-made materials that are suitable for implantation, and new ways to produce materials with biological compatibility.

Here's an example of a highly successful utility patent. Two years after forming the ALZA Corporation, Dr. Alejandro Zaffaroni filed a patent application that was entitled "Drug-Delivery System" in June, 1970. The application comprises a porous reservoir that contains a drug for delivery to a host environment at a controlled rate for a prolonged period of time in order to produce a local or systemic physiological or pharmacological effect. As the first of many such patents issued to ALZA, the USPTO granted the patent in December, 1974. Built on the development and marketing of this and many other drug delivery systems, the ALZA Corporation has become a billion-dollar company and a leader of drug delivery solutions by oral, transdermal, implantable and liposomal technologies. The corporation has approximately 3,000 active or pending US and foreign patents. It is interesting to note that Zaffaroni is the inventor or co-inventor of more than one hundred patents, the latest of which was granted in 2008.

Design Patents. Design patents, in essence, are given to new, original and ornamental designs of articles of manufacture. Design patents protect only the appearance of a manufactured article and not its structural or functional features.

For example, when you manufacture the central processing unit (CPU) of a computer, you lay out the semi-conductors or transistors in a particular design that processes and transmits various binary signals. The circuit design of your CPU is patentable. That is why, whenever Intel came up with a new Pentium CPU, the company would apply for a design patent. However, if Intel invented a new way of packing transistors into a CPU to achieve a much higher processing speed, it would file a utility patent, which covers matters concerning the functionality of the CPU.

Plant Patents. Plant patents may be granted to anyone who invents or discovers and asexually reproduces a distinct and new variety of plant. If you insert a gene into a plant so that it will produce some special chemical or pharmaceutical substance, you may have a patentable plant.

Plant patents also apply to innovations that involve cells, animals, or other biological entities.

Additional Patenting Considerations. Each of these three patent types is screened according to three criteria: new, useful, and non- obvious. For example, if you come up with an invention, design or plant, but patents for similar inventions, designs or plants already exist, you cannot patent your invention. An idea becomes patentable only when you generate or produce new versions of existing inventions. Similarly, if a medical device is not deemed to be useful, you cannot patent it. For example, a thermometer that does not achieve its intended purpose, the measurement of temperature, is not patentable. Also, you cannot patent a process or device if it is obvious that everyone already knows about it. The term "non-obvious" or "sufficiently different" is determined by a "person" who possesses skill in an area of technology that is related to the process or device.

Patent requirements and the impact of being granted a patent are illustrated in Fig. 7.1.

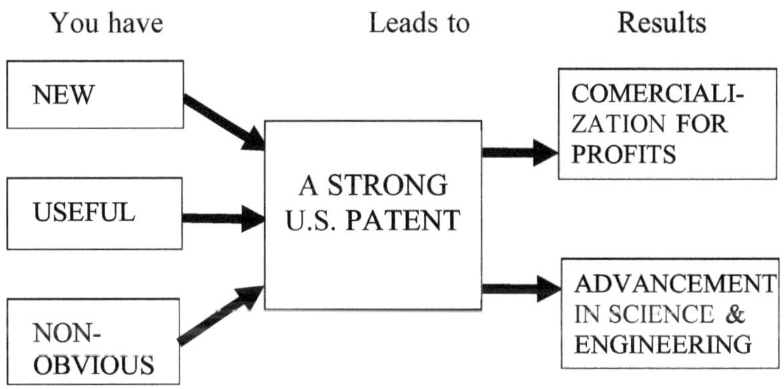

Fig. 7.1. The features of a strong US patent and the impact on the commercialization of the product and the advancement of science and engineering.

If your patentable (but currently unpatented) invention is published in public media, such as newspapers and journals, or presented in public forums, such as conferences or classrooms, it no longer can be patented. If you want to publish your results before a patent has been filed, you can file a provisional patent with the USPTO. The provisional patent allows you to publish your results after the filing while you still have a year in which to submit your patent application.

Within a year of receiving a patent approval, you may submit a continuation-in-part for any new invention that is an improvement or redesign of the original patent. The attorney costs of filing a continuation-in-part are lower than the cost of the initial patent, which provides sufficient reason to file one, when appropriate. Another reason to file a continuation-in-part within the first year is that applications that are submitted after the first year of the initial patent approval may be denied by the USPTO, due to similarities between the original patented product and the one for which an application for the continuation-in-part was submitted.

Patents have 20-year lifespans. Even if you file a provisional patent application before the non-provisional application, the 20-year term of the patent begins on the date on which the non-provisional patent application was received by the USPTO. If it takes two years for USPTO approval of your non- provisional patent application, you will have 18 years remaining on your patent in which to complete the development of your invention and manufacture and sell it without competition in the United States.

Until your patent expires, you must pay an annual fee to the USPTO to maintain the active status of your patent. Fees are lower for patents that are granted to small businesses.

7.3. Trademarks and Copyrights

Trademarks and copyrights are two other types of legal protection that the federal government grants to intellectual properties. Trademarks protect words, names, symbols, sounds or colors that distinguish goods and services. If the words that are used in your trademark describe a general class of goods or one with a specific, but general, function, the USPTO may not grant you the trademark.

Copyrights protect works of authorship that have been tangibly expressed. They apply to books, music CDs, movies and software.

Although all patents, trademarks and copyrights have a defined period of protection, Congress has voted on occasion to extend the period for some trademarks and copyrights. In general, however, trademarks that were registered after November, 16, 1989, are effective for ten years, but, unlike patents, can be renewed forever by the USPTO as long as the trademark continues to be used in business. Copyrights are registered by the U.S. Library of Congress and last for the lifetime of the author, plus 70 years.

It is worth noting that there is a major difference between the copyright of a book and that of a piece of computer software. A book can be lent to many readers without violating its copyright, but software cannot. Computer software can be loaded onto one computer for use by one or more persons, but not onto several computers for use by many people. This practice extends to the health care industry in that a copyrighted software can only be installed in one particular machine for diagnosis and treatment. However, this machine can be used by many users for many patients.

A software licensing agreement can provide for the rental of the software. In other words, the user will pay a rental fee whenever the software is used.

Your software should be copyrighted. You can upgrade the software to increase its power in processing, diagnosis or treatment. This is another avenue for your company to continuously generate new revenues and profits.

7.4. The Importance of Patenting Your Intellectual Property

The disclosure of your invention can lead to advancement in the nation's science and technology. It is equally important for the prosperity of your company that you file a patent application so that you can obtain patent protection for your invention to maximize its profitability. The first concern is the expense of obtaining patent protection, especially for a start-up that has limited capital. The second concern is that your disclosure may enable your competitor to invent an even more efficient and cost-effective product that can put your company out of business. In the following pages, we will consider a number of issues to help you decide whether patent protection will overcome these two concerns.

An alternative to patenting is to decide not to disclose your invention, but to protect it as proprietary information. You need to ensure that your invention cannot be duplicated by reverse engineering. Date all documents that are related to your proprietary information for use in a court of law in case your ownership of the proprietary information is challenged by your competitors holding other patents.

Suppose that someone filed a patent at the USPTO well before you or obtained a patent for an invention that is similar to yours. As you introduce your product to the market, they can bring suit in US Courts for an injunction to stop the sale of your product because it infringes on their patent. When the sales of your product begin to develop significant revenue, they can also file an infringement suit and request damages and a punitive payment. Long before the case ends, you will incur large attorney fees to defend your case.

In such a legal proceeding, whether your competitor has its product on the market is not relevant if it is diligently developing the product. See the discussion on patent troll in Section 14.1. Your competitor's patent readily puts it into a position to sue you for patent infringement. If you were to lose the suit, you could no longer sell your product and your company might be obligated to pay huge punitive damages. Such a legal challenge, whether successful or not, would impact your company financially.

It is a well-known fact that the Coca-Cola Company owns the world's most famous commercial trade secret, the formula used to make Coca-Cola, a product invented in 1886. Asa Candler bought the formula and brand in 1886 and incorporated The Coca-Cola Company in 1889. The company's 2016 revenue was about $42 billion and its net income was $8.6 billion. No patent was ever filed for the formula. The company's current policy still is to keep most of its product formulations as trade secrets. Don't use this as an excuse to not file your own patent application as your medical invention is very different from a beverage formula. Your invention may be easily produced with the "same" function, whereas a new beverage can only be imitated because the "taste" may still differ. Secondly, please note that the patent position of Coca-Cola in beverage technologies is strong. The Coca-Cola Company has approximately 800 US patents and 1,800 international patents (2).

During the development of your company, you will approach venture capitalists for funding. The question of your patent situation and means to prevent competitors from entering the market will always come up in meetings with them. A poor response to the question about patent protection will often mean that you will receive no venture capital funding.

There are many successful companies that have strong patent portfolios. In 1997, Boston Scientific acquired Target Therapeutics in a tax-free stock swap for approximately $1.1 billion (5). This sale price was more than 10 times Target's total sales, in contrast to the usual formula of ten times earnings. The reason for this high price was that Target had pioneered the field of interventional neuroradiology and neuroendovascular therapy and had dominant patent positions in this area by having been the first mover in certain stent developments. Their Guglielmi Detachable Coil for the treatment of inoperable or high risk brain aneurysms was the first device in that category to be approved by the FDA. Guido Guglielmi and Ivan Sepetka are the inventors listed in the USPTO patent #5,122,136, which was filed in March 1990 and granted in

June 1992, entitled "Endovascular Electrolytically Detachable Guidewire Tip for the Electroformation of Thrombus in Arteries, Veins, Aneurysms, Vascular Malformations and Arteriovenous Fistulas". At the time of acquisition by Boston Scientific, Target was the plaintiff in litigation with Scimed (a company acquired by Boston Scientific in 1995) and Cordis (a Johnson & Johnson company) over a microcatheter.

Suppose that the USPTO grants you a patent. This is a grant of property rights to the inventors, or their assignee, to exclude others from making, offering for sale or selling the invention in the US or importing the invention to the US. You must enforce the rights afforded to you by your patent by taking legal actions in US Courts without any aid from the USPTO. If you win the patent infringement case, you may be entitled to lost royalties and, in some cases, to punitive damages. Because of the significant effort spent on the discovery of evidence of infringement, processing an infringement case may incur legal fees of millions of dollars. There is more discussion of patent lawsuits in Section 14.1.

7.5. The Process of Patenting

Once you have an invention, decide if it is patentable. Will it be considered to be new, useful, and non-obvious? Search the public records to learn if there is a device that is similar to yours. In some cases, before a non-provisional patent is approved, the inventor and the USPTO may keep the patent application confidential for a period of time. There is also the matter of finding foreign patents by means of the Internet. Suppose that you find devices on the market that are similar to yours. If the difference between your device and those that you have found is "non-obvious," you can proceed to file an application with the USPTO for a provisional or non-provisional patent.

If you consider that the invention has commercial potential and your small business has $199 available for the filing fee, you can apply by yourself electronically or by mail for a provisional patent application. It consists of the following parts:

- Provisional Application for Patent Cover Sheet
- Application Data Sheet 37 CFR) 1.76
- Drawings of your invention
- General description of your invention

Claims and an oath or declaration are NOT required for a provisional application. Later, you will receive a letter from the USPTO that indicates that an application number has been assigned for your provisional patent application and the filing date. The letter will also advise that the provisional application will NOT be examined by the USPTO for patentability. You will then have up to 12 months to file a non-provisional application. If you fail to do so, the provisional patent application is considered to have been abandoned. Because the provisional patent application is published, its abandonment can be used by the USPTO as a challenge to the patentability of your future patents of device of a similar nature.

One advantage of filing a provisional patent application is that you are immediately entitled to label your device as "patent pending." As soon as the non-provisional patent has been granted, you will have the right to take legal action for any violations of your patent rights that occurred after the filing date of the provisional patent application.

You can file a non-provisional patent without filing a provisional patent. The key advantage of filing a provisional application first is that will have the benefit of an earlier filing date.

If you are 65 years of age or older, you can file a petition to have an earlier examination of your application due to your age. If it is granted, the USPTO examination process will be speeded up.

Before 2013 the United States was the only country in the world that awards a patent to whomever is the first to invent the device, whereas other countries award a patent to the inventor who is first to file a patent application for the invention. US switched to a first-inventor-to-file system after the enactment of the America Invents Act.

The legal protection afforded to a patent that has been approved by the USPTO applies only to the United States market and its territories. The global market position of your medical inventions may warrant your filing a patent internationally, as well. See Section 18.4 for more information on foreign patent filing.

7.6. Key Elements of a Non-provisional Patent Application

The information required for a patent application is pretty straightforward. The application contains the following sections:

- The title

- Cross-references to related applications
- Field of subject matter or technology
- Background
- Summary
- Brief description of the drawings
- Detailed description
- Claims
- Abstract
- Figures

The application for a patent involves a number of forms and is submitted to the USPTO for patent examination. The sections listed above, albeit in a slightly different order, together with the name of the inventor and publication classification, will appear in the public records of the USPTO in about one year from the time of filing.

The background of your invention must be thorough and accurate, because it will be the first item to be reviewed by the patent examiner. This is where you provide the patent examiner with an assessment of what is currently available in the market. Good research is required to support the accuracy of your descriptions, as well as reporting any comparable patents or similar devices of prior arts. The discovery of an omission of information or over-inclusion of materials in your patent search may be used by the USPTO as a reason to invalidate your patent, even years after it has been granted.

If your patent application covers two or more inventions that are somewhat independent and distinct, the USPTO will return you application and require that you limit the application to a single invention.

The detailed description of the invention needs to be as specific as possible. If the dosage of the drug, the operational range of the device, the composition of the materials or some other factor is essential for the use of the machine or process for which the patent is being sought, it should be spelled out in the description.

The "Claims" section is where the inventor claims what his or her device, material, drug, design, or plant does. It is good to make your claims as broad as possible, when this is appropriate.

You also must describe the purpose of the invention. In the case of a medical device, you would indicate the medical problem that you intend to solve and how this differs from what is described in your competitors' patents. If your device has technical characteristics that are similar to those of your competitor, but your application is for a different disease, it is likely that your device will be perceived to have a new use and thus be patentable for that disease.

A series of drawings and/or examples of the device's use in the description section provides a clear explanation of your invention. The drawings should include illustrations of the invention, and the experimental data and theoretical analysis that are critical to the invention's application.

Finally, you should describe the special features of the invention in the context of preferred embodiments. The USPTO's rules also make it clear that you do not need to have a workable prototype in order for you to file your patent application.

Who should prepare the patent application? In general, someone who is knowledgeable in research writing should be able to prepare the description part of the application. On the other hand, the claims section of the patent, which can contain many legal terms, may be best left to an experienced patent attorney. A patent attorney will help you to write a clear and concise patent document that will not only enable you to obtain the approval that you seek from the USPTO, but also will eliminate any ambiguities in the application that could be detrimental in the enforcement of your patent in future infringement lawsuits.

The attorney's fees for a non-provisional patent application can range from several thousands of dollars to tens of thousands dollars. This is much higher than the expense of filing a provisional patent application. As with everything else, you get what you pay for. Make sure that your patent attorney has experience in the scientific and engineering field of invention and that you put in the effort and funds to help your attorney do the job right.

7.7. Special Patenting Considerations

Orphan Drugs. Many diseases occur only in a small population. The prospect of limited sales may discourage pharmaceutical companies from developing new drugs for certain diseases. The reason is drug development is an expensive proposition with considerable uncertainty about whether the drug that is being developed will be able to control or cure these diseases. Drugs that are developed for diseases that affect small populations are commonly referred to as orphanage drugs. In order to encourage their development, a law was enacted that grants a patent for the development of orphan drugs, which has a life that is longer than the conventional 20 years.

Criticism or Rejection of a Patentable Idea. Suppose that your attorney advises you that your invention may not be new, useful, or non-obvious and, therefore, filing an application for a patent will probably be a waste of your time and money. Alternatively, your attorney may point out that your patent would infringe on a number of existing patents. Don't be discouraged. Instead, view these criticisms as an incentive for you to improve your patentable idea.

If you proceed, the USPTO will generally notify you in writing of the patent examiner's decision within one or two years. If your patent application is rejected, you can either request reconsideration by pointing out perceived errors in the examiner's decision, or significantly amending your application and explaining why the amended claim is now patentable.

If the amended application is rejected by the patent examiner, you may have little further recourse. An appeal of the examiner's final decision is possible. However, it is greatly restricted in the scope of amendments.

Gene Mutations and Gene Patenting. Because family history is the strongest single predictor of a woman's chance of developing breast cancer, researchers turned to cancer-prone families - those with a high incidence of cancer in several generations – and used big data analysis to compare the genes of these patients to identify specific inherited gene alterations that are passed on from one generation to the next. After a long search, the breast cancer gene BRAC1 was found by Mary C. King, professor of genome sciences and of medicine at University of Washington in 1994 and the second gene BRCA2 in 1995.

In 2013 the Supreme Court ruled that companies cannot patent isolated human genes that are naturally occurring. The Supreme Court also ruled that synthetically-created DNA (known as cDNA) is patentable since it is not naturally occurring. This decision has the following four impacts (6):

1. **There will be competition in BRCA testing.** Since physicians will not be compelled to use a particular test, the competition in genetic test will help control prices. This decision will allow BRCA to be incorporated in a panel of genes that physicians may want to evaluate and test patients for.
2. **Product development will speed up.** The decision removes the threat of endless litigation if a product–for instance a screening

method–is developed by one company that uses a gene patented by another.

3. **Biotech firms will have more freedom to innovate.** This means that potential competitive products cannot easily be quashed and the flow of research in all sectors–academic, industrial and clinical–will continue unimpeded.
4. **There will be more personalized therapies that rely on gene sequences.** If the natural gene was patented, the development of personalized therapies would not be commercially possible.

Patent Policies of Your Employer. Suppose you work in a research company or university, and invent something in your garage in your spare time without using any of your employer's materials or equipment. Can your employer claim ownership of your invention, or can you obtain a patent for yourself and then build up a company on the order of what Drs. Bill Hewlett and Dave Packard achieved? The answer depends strictly on your employer's patent policy concerning employee inventions.

Read your employer's patent policy carefully before accepting his offer of employment, especially if signing an employee agreement is a prerequisite to your working for the firm. Some companies allow you to invent things in your own garage and to own your patents if they are not related to your work in the company. Other companies stipulate that they legally own everything that you invent at work or elsewhere.

Strong Patent Operation within. Many high-tech and software companies have their own patent offices. Besides helping their companies to prepare patent application, they spend significant effort in defending patent lawsuits or pursuing actions against competitors for patent infringement. The annual budget of such patent departments can be as high as hundreds of million dollars.

For many start-ups in the medical device and biotech field, patents are the key property that they own when valuing their company. If you have an invention with a strong patent of broad claims, your company will have a strong competitive edge that may lead to capturing a large share of the market and, consequently, a large profit from sales of the device. Because of the potential of such a patent to raise venture investment, many small companies invest heavily in their patent operation and pursue infringement lawsuits aggressively.

Ten Common Mistakes to Avoid. An excellent analysis by Michael Neustell of how to avoid making common patenting mistakes appeared in IEEE-USA Today's Engineer (4). Here is a summary of the common mistakes:

- No employee education
- "Can't be patented" attitude
- No intellectual property management program
- Conducting patent research too late
- Failure to perform IP audits
- Not taking advantage of "provisional" applications
- Filing the application at the wrong time
- Patenting the wrong inventions
- Failing to evaluate competitors' patent rights
- Filing unnecessary foreign patent applications

Patent Litigations. Erythropoietin (EPO) is the principal hormone involved in the regulation and maintenance of a physiological level of circulating red blood cells. In two 1985 publications, Fu-Kuen Lin and his coworkers of Amgen described a means of producing a sufficient quantity of EPO for treatment of anemia and the gene expression of the cloned EPO (3). Subsequently Amgen filed a patent application on Nov 30, 1984 and obtained United States patent number 4,703,008 for a process to obtain erythropoietin on 27 October 1987. Earlier Genetics Institute filed an application relating to a process for purifying EPO on January 11 1985 and obtained United States patent 4,677,195.

These two patents resulted in an intense legal battle over who owns the rights to the genetic code sequence of EPO and the method of obtaining EPO. At that time, there was significant debate on whether a substance occurring in nature can be patented. In addition, the method of obtaining proteins by genetic manipulation was already widely known. In 1996, the United States Court of Appeal ruled that the rights of Amgen Inc. overrode those of Genetics Institute.

This ruling was sustained by Judge Lourie with this reasoning: priority of invention went to the first person who could demonstrate 'the structure or physical characteristics' of the EPO gene and "had a viable method of obtaining it.' In his opinion, the gene's identification by its isolation and the sequencing of its DNA were the inventive steps – without these steps being taken the protein EPO, which the gene coded for, could not be synthesized." This led Judge Lourie to hold that "the trial court was correct in concluding that neither party had an adequate conception of the DNA sequence until reduction to practice had been achieved; Lin was first to accomplish that goal."

By vindicating the patent on EPO in many countries and completing its clinical trial, only EPO that is produced by Amgen can be marketed. In 2003, Amgen sold $2.4 billion of its Epogen treatment worldwide.

60% of Genetics Institute was acquired by American Home Products[a] for $666 million in 1991 and the remaining 40% for $1.25 billion in 1992. The May 2017 valuation of Amgen by Forbes is $120 billion.

Value of a Patent. An approved patent is an asset for your company. As discussed previously, the EPO patent is certainly important to the marketing position of Amgen and the Apple's patents to setting the amount of damages due to infringement as examined next in this subsection. However determining the value of a patent is still a subjective exercise. Just like setting the market value of houses or companies, the final determination of a patent's value will be determined only when the investors and the owners of the company agree on the patent's value. This point is illustrated by the following two patent acquisitions.

First a group of high tech heavyweights, including Apple, Google and Facebook, have joined forces to purchase approximately 1,100 digital imaging and processing patents from Eastman Kodak for $525 million. Second, Google reported in 2014 that it retained all the patents and applications of Motorola Mobility for less than a billion dollars after Google sold Motorola for around $3 billion to China's Lenovo. (Google acquired Motorola for $12.5 billion in 2011.) Google's CEO Larry Page said, "Google will retain the vast majority of Motorola's patents, which we will continue to use to defend the entire Android ecosystem."

In patent litigation of Apple against Samsung, a California jury ruled in May 2014 that Samsung had violated two of five software Apple patents. Additionally, it ruled that Apple had violated one of two Samsung patents, but not willfully. Apple had been seeking $2.2 billion in damages, but the jury awarded just $119.6 million. It also awarded Samsung $158,400. (Two years before, Apple was awarded a more imposing $930 million over a different set of design-related patents.)

[a] American Home Products was renamed as Wyeth Pharmaceutical in 2002 and acquired by Pfizer in 2009.

These Apple Samsung sagas are some of the patent wars that are fought on the operating system (OS) of Google and Apple. The fight ended two weeks after the 2014 verdict on Samsung's patent violation. They stated further that "Apple and Google had also agreed to work together in some areas of patent reform and that their deal did not include an agreement to license technology to each other."

For 2016, 86% of the world's smart phones employed the Android OS and 13% used the Apple OS. In discussing this truce, "It's a huge relief to see these companies finally recognize that competing in the courtroom doesn't make sense. Instead of spending millions of dollars enriching patent lawyers, they'll put those resources into real competition — in the marketplace." said Julie Samuels, a top intellectual property lawyer who served as a staff attorney at the Electronic Frontier Foundation and who is now executive director of Engine Advocacy, a nonprofit tech policy group (1).

References

1. Gustin S., Apple and Google Call a Truce in Patent Wars. *TIME*, May 17, 2014.
2. Landgrave F.A., Patenting at Coca-Cola, www.yet2.com/app/insight/insight/20010401_landgraff (2001).
3. Lin F.K. et al., Cloning and expression of the human erythropoietin gene, *Proc. Natl. Acad. Sci.* 82:7580-7584, (1985).
4. Neustel M.S., Patents: 10 Common Mistakes and How to Avoid Making Them, www.todaysengineer.org/2008/apr/patents.asp (2008).
5. Rodengen J.L., *The Ship in the Balloon*, Write Stuff Enterprises, p. 173 (2001).
6. Sifferlin A., 4 ways the supreme court gene patent decision will change medicine, *Time*, June 14, 2013 healthland.time.com/2013/06/14/4-ways-the-supreme-court-gene-patent-decision-will-change-medicine/

Chapter 8

FDA Regulations

8.1. The Food and Drug Administration

What does the FDA regulate? The Food and Drug Administration, or the FDA, was authorized by the Federal Food, Drug and Cosmetic Act in 1938 to regulate products that are covered by the Act's name, although its scope has grown through the following years.

In 1968, for instance, the Radiation Control for Health and Safety Act brought some medical devices under the oversight of the federal government. The Medical Device Amendments of 1976 were passed to ensure the safety and effectiveness of all medical devices, including diagnostic products. As a result of these and other legislative charges, the following product categories are regulated by the FDA today, according to the agency's website (www.fda.gov):

- Food
- Drugs
- Medical devices
- Biologics
- Animal feed and drugs
- Cosmetics
- Radiation-emitting products.

In addition to the legislation mentioned above, other Acts of interest to entrepreneurs in the field of biomedical engineering include the Safe Medical Devices Act of 1990. The latter requires manufacturers to conduct post-market surveillance of permanently-implanted devices whose failure might cause serious harm or death to a patient. Another Act of interest to entrepreneurs in the biomedical engineering field is the FDA Modernization Act of 1997. Among other items, it contains measures to speed up the approval process of medical devices. Product user fees were also implemented.

FDA Research Tools. Federal government regulations are first published in the Federal Register under Code of Federal Regulations (CFR). Title 21 CFR is reserved for FDA regulations. The section that deals specifically with medical devices is codified in Title 21 CFR Part 800-1200.

The FDA also has a set of Compliance Policy Guides (CPG) for each of the product categories listed earlier. The guides for medical devices are provided in Chapter 3 of the CPG.

The two documents mentioned above are available online from FDA websites. They are the research tools that are necessary to learn about the FDA approval process for your particular invention.

In addition, the FDA's Center for Devices and Radiological Health (CDRH) has a website document called "Device Advice." It provides a comprehensive description of the rules, regulations, and processes that are important in seeking approval of your device, as well as in manufacturing and marketing it.

The following sections contain information about other important issues that you will need to keep in mind or pursue while you engage in the research and development, manufacturing, and marketing of your device. Since many medical device companies will be submitting a Traditional 510(k) Application, Chapter 9 discusses this topic.

8.2. The Safety and Effectiveness of Medical Devices

Part of FDA's mission is to reasonably ensure the safety and effectiveness of devices designed for human use. Some FDA regulations that are intended to support this mission are fairly broad, whereas others are quite specific.

The Radiation Control for Health and Safety Act, for example, contains a general paragraph that, in essence, says that the FDA should not allow the US public to be exposed to radiation, as radiation is harmful. The Safe Medical Devices Act, on the other hand, spells out very specifically who can serve on a medical device advisory panel. However, the US Congress has given the FDA a fair amount of discretion to write regulations that are in between these two poles of generality and specificity, based on the latter's interpretation of the law. How does the FDA determine what is safe and effective?

One way is by relying on valid scientific evidences. A device manufacturer needs to carry out investigations using laboratory animals, conduct certain investigations involving human subjects or undertake in vitro studies to demonstrate that the use of a medical device outweighs any probable risks and imposes no unreasonable risk of illness or injury. When well-controlled investigations are conducted within a substantial portion of the target population, clinically significant results from the investigation are taken to mean that the use of the device is effective.

In determining safety and effectiveness, the FDA also considers factors such as:

- The persons for whom the device is intended,
- The conditions under which the device is used,
- The probable benefit to health, and
- The device's reliability.

Other considerations may include the following:

- Will the device be used for children?
- Will it be used in a hospital?
- Is it a prescription device or will it be sold over the counter?

The answers to these or other similar types of questions will determine the device's classification and how its safety and effectiveness will be evaluated.

8.3. Device Classification and Indications for Use

What is a Medical Device? By law, its definition is broad, with one part of the definition considering it as a physical device, and another part considering its intended use. Consequently, the FDA may classify it as an instrument, apparatus, implement, machine, contrivance, part, accessory or as a medical device that is intended for use in the diagnosis of disease, treatment of disease or functional improvement of the body. For example, the Kolthoff blood pressure cuff to measure blood pressure non-invasively is a medical device, whereas a weighing scale a nurse uses to determine the obesity of a patient is a consumer product that the FDA may decline to regulate.

Every medical device is classified in one of 16 medical specialties, such as Cardiovascular Devices, Neurological Devices, and Orthopedic Devices. These medical specialties are further divided into subcategories. For example, the 137 generic types of devices that are found in the Cardiovascular Device medical specialty fall into one of five subcategories, which are Diagnostic, Monitoring, Prosthetic, Surgical and

Therapeutic Devices. Overall, approximately 1,700 different generic types of medical devices fall into these various specialties and subcategories.

The FDA also classifies medical devices as Class I — General Controls, Class II — Special Controls and Class III — Pre-Market Approval (PMA). These three classifications are based on risk, with Class I posing the lowest risk to the patient and/or user and Class III posing the greatest risk.

Class I Devices. These devices are subject to the fewest regulatory controls. Examples of Class I devices include stethoscopes, elastic bandages, examination gloves and surgical scissors. In keeping with the commitment of the FDA to speed up patient access to new medical devices, the agency exempted 122 generic types of Class I devices from 510(k) submission in 1996. These newly-exempted devices include oxygen masks, pacemaker chargers, dental floss, patient scales, examination lights and therapeutic massagers. This FDA action made approximately 572 (or 74%) of Class I generic types exempt from 510(k) of the Food, Drug and Cosmetic Act, which requires device manufacturers to notify the FDA of their intent to market a medical device.

Class II Devices. These devices require more consideration by the FDA than Class I devices to ensure safety and effectiveness. A 510(k) submission is required for most Class II devices in order for the applicant to establish that the device is substantially equivalent to a legally marketed device. Electronic stethoscopes, blood pressure cuffs, infusion pumps, steerable catheters, cutaneous electrodes, surgical drapes and powered wheelchairs are considered to be Class II devices. With the implementation of the Modernization Act, the FDA reclassified 111 generic types of Class II devices as Class I devices.

Class III Devices. These devices are generally those that support human life or prevent impairment of health. They require the premarket approval (PMA) to ensure the safety and effectiveness of devices before they can be used. Class III devices include implants, such as metal knee joints, replacement heart valves, silicone gel-filled breast implants and cerebella stimulators.

If your Class III device was available in the market prior to the adoption of the Medical Device Amendments in 1976 or substantially equivalent to such a device, you may want to consider using the 510(k) process as a better route to the market, because it is easier and faster than the PMA process.

Think of it this way. While the NIH will be interested in funding your research and development of a medical device, even if it is an improvement of some existing device, you may still want to market your device by advertising it as new and totally novel. To do so, you must go through the PMA process. This is laborious and time-consuming. You, as the manufacturer, may face a much greater financial burden to get it to market and pay more in user fees and consultant costs.

Even if you think that your device may provide a more accurate diagnosis, but you can demonstrate that it is substantially equivalent in terms of safety and effectiveness to other Class I or Class II devices on the market, the 510(k) process is the one to use.

If the FDA finds that your device is equivalent to a Class III device, and you believe that the device is more appropriately a Class I or II device, you can petition the FDA to reclassify the device to a lower class. Anyone can file this type of petition, even someone who is not marketing the device. If the petition is successful, it will no longer be necessary to go through the PMA process for the Class III device.

According to the 510(k) policy unit within the CDRH's Office of Device Evaluation, 90% of devices are cleared by 510(k) with others either going through the more rigorous PMA path or being deemed 510(k)-exempt. Clinical trials are required for about 10% of 510(k)s.

8.4. Investigational Device Exemption (IDE) and Clinical Trials

When you are developing or modifying a medical device, your business may need to perform clinical or animal trials to obtain data that demonstrates the safety and effectiveness of the device. An approved Investigational Device Exemption (IDE) permits a device that has not received 510(k) approval or PMA to be tested for human use. Without such facilities, the small business needs to work out a contractual arrangement with the teaching and research hospital of a university by which they will become a sponsor for the execution of the trials. The sponsor will be responsible for selecting investigators, usually faculty members, for the trials. The investigators will submit a research protocol and an assurance of humane care and use of laboratory animals for approval by the university's Institutional Review Board (IRB) Only with formal approval by this Board can the company and the investigators carry out the animal experimentations and/or clinical trials.

If you are going to use human subjects to develop the data needed by the FDA for approval or to market the device, you must first decide if there is a significant risk to patients in a trial of this device. If the trial poses insignificant risk, the investigator will submit the research protocol and patient consent forms to the IRB for review. If the Board grants you an IDE for your device, you can begin your trial. When a device is considered by the IRB to involve risk to patients, the sponsor of the clinical trial must obtain an IDE from the FDA before beginning any clinical trial of the device.

As an alternative, you may delegate the clinical trials entirely to a contract research or services organization.

An application for an IDE from the FDA contains several items, including a description of prior investigations of the device, a description of the methods, facilities and controls used in manufacturing the device, a signed agreement by all investigators to comply with their investigator's obligations, a list of the IRB(s) that have been or will be asked to review the investigation and a plan of investigation. The plan of investigation includes a description of the device, a description of the proposed use of the device, a clinical protocol in order to obtain a reasonable assurance of effectiveness, informed consent material and, if available, information about the device's expected performance. The FDA will act on your exemption request within 30 days.

The law states that you can only ship FDA-approved devices within the United States. Upon receipt of an IDE from FDA, the shipment of testing devices to institutions for clinical trials becomes legal.

However, the primary rationale for the IDE process is to assure that the rights, safety and welfare of all research subjects are protected. This is an important consideration. It was legislated by the Safe Medical Devices Act of 1990 to avoid the abuse of patients in clinical investigations.

When Barney Clark received an artificial heart in 1982, the artificial heart (known as the Jarvis 7) had been implanted previously only in calves to test its performance. At this experimental stage, there were no data that would indicate how long a patient who had an artificial heart would live and how risky the implant would be to the patient. However, Clark, who understood the risks of artificial heart implantation, the probable future benefit to other humans if the implantation worked and who wanted to do something altruistic for his fellow men, was, accepted as a volunteer for the clinical trial because the Jarvis 7 had received an IDE from the FDA.

To ensure that a patient fully understands his or her rights, the patient consent form describes his or her role in the clinical trial, the risks posed by the trial and the countermeasures and protection that the clinical trial's physicians and investigators will take. All statements are in layman's terms. The patient consent form also stipulates that the patient may withdraw from the trial whenever he or she chooses to do so.

8.5. Premarket Approval Applications

If your performance data show that your device is, to a large extent, equivalent to a Class III device without exemption, you must submit a PMA) application before you market your device. Usually the Class III device is something for implantation in a human body. This means that you will need to obtain clinical data in order to demonstrate that every part of your device is safe and effective. When you invent a new material, you must first show with appropriate animal implantation that your new material is suitable for implantation in the human body. With this animal study, you can obtain an IDE from the FDA so that you can begin clinical trials. You must carry out a full scientific investigation that will support your PMA application. Although this requires a great deal of work, the FDA has much information available to help you to prepare the PMA application.

The FDA takes a minimum of 180 days to reply to a PMA application, which will be reviewed by an advisory panel of medical device experts. The panel may skip a full review if it believes that similar implantable materials have been reviewed and approved by its members previously. Finally, the advisory panel will make a recommendation to the FDA to approve or not approve the application. The FDA nearly always accepts the advisory panel's recommendation.

8.6. Artificial Pancreas

In September 2016, the FDA approved Medtronic's artificial pancreas. The PMA application had been submitted to FDA three months previously. The system is the first FDA-approved device that continuously measures glucose levels and delivers the appropriate dose of basal insulin by an adaptive control algorithm.

The system comprises Medtronic's MiniMed 670G insulin pump that is strapped to the body, an infusion patch that delivers insulin via a catheter from the pump and a sensor that measures glucose levels under the skin and can be worn for seven days at a time. Although the device regulates basal, or background, insulin, users must still manually request bolus insulin at mealtimes.

The approved indications for use are described in the following paragraphs: "The MiniMed 530G System is intended for continuous delivery of basal insulin (at user selectable rates) and administration of insulin boluses (in user selectable amounts) for the management of diabetes mellitus in persons, sixteen years of age and older, requiring insulin as well as for the continuous monitoring and trending of glucose levels in the fluid under the skin."

The FDA assessed data from a clinical trial in which 123 patients with Type 1 diabetes (T1D) used the system's hybrid closed-loop feature as often as possible during a three-month period. The trial showed that the device was safe for use in those of age 14 and older as no serious adverse events were reported.

In conjunction with this FDA announcement, Dr. Jeffrey Shuren, director of the FDA's Center for Devices and Radiological Health said, "This first-of-its-kind technology can provide people with T1D greater freedom to live their lives without having to consistently and manually monitor baseline glucose levels and administer insulin."

As many as three million Americans have T1D, according to the Juvenile Diabetes Research Foundation (JDRF). About 15,000 children and 15,000 adults are diagnosed each year. People with the disease must carefully balance insulin doses with eating and other activities throughout the day and night. They must also measure their blood-glucose level by pricking their fingers for blood six or more times a day. The T1D associated annual healthcare cost in the U.S. is $14 billion.

An artificial pancreas that was developed at the University of Cambridge and given to children and adults with T1D going about their daily lives has been proven to work for 12 weeks. This means that the technology can now offer an entire school term of extra freedom for children with T1D. This 2015 Cambridge study showed that the artificial pancreas improved control of blood glucose levels among participants–lessening their risk of hypoglycemia. Known as 'having a hypo', hypoglycemia is a drop in blood glucose levels that can be highly dangerous and is what T1D patients hate most.

In August 2017, Tandem Diabetes Care announced that the FDA had approved its G5 integration. This is the California Company's fifth iteration to its touchscreen insulin pump since first hitting the US market in August 2012. This system employs the Dexcom CGM for continuous glucose monitoring. Tandem believes that this devices has these important and influential changes, the first G5-integrated pump available in the U.S., the first one officially approved for dosing and treatment decisions, and the first model with the latest features that can be remotely updated from home.

In October 2017 the FDA and Abbott announced the US approval of a real-time "continuous glucose monitoring system". for use by adults with diabetes only. This "FreeStyle Libre" device does not require any daily fingerstick calibrations and real-time readings and trends can be used for insulin dosing. The average measurement error relative to a reference value is 9.7% and on par with 9.0% for Dexcom's G5 and with 10.6% for Medtronic's Guardian Sensor 3. FreeStyle Libre has a 12-hour warmup, whereas Dexcom and Medtronic have a two-hour warmup period.

Abbott is continuing to evolve FreeStyle Libre, including new product features and indications. The US continuous glucose monitor market for FreeStyle Libre will expand because of the lower price, no need for fingerstick calibration, small on-body footprint, ease of insertion, and pharmacy distribution. This device has been available in Europe since fall 2014. People with diabetes globally have had a very positive response to this device, and it's now used by more than 400,000 people across over 41 countries.

8.7. Post Approval and Post Marketing Requirements

Post-approval. Manufacturers of Class I devices with exemption must follow general control procedures for the following:

- Registering their company with the FDA as a device manufacturer,
- Listing the device being marketed with the FDA,
- Implementing FDA-defined Good Manufacturing Practices[a].
- This includes documenting how the device is made, ensuring that the labeling of their device is truthful and that they have developed adequate instructions for its use.

---

[a] As set forth in the Quality System regulation promulgated under Section 520 of the food, Drug, and Cosmetic Act.

If no adverse effect from use of your Class I device is reported, no post-marketing requirements may be necessary. The factory will probably not require an inspection by the FDA for a Class I device, unless there is a complaint about the device and the FDA feels compelled to follow up on the complaint with an on-site inspection.

In order to receive post-approval market clearance for Class II devices submitted through the 510(k) process, you must not only follow the general control procedure required for Class I devices, but also ensure that your Class II device follows the special labeling requirements and satisfies the mandatory performance standards. Typically, the facility in which you manufacture your Class II device must be inspected every two years. During the first inspection, the FDA inspector will examine performance data to make sure that what you stipulated in your 510(k) application is applicable to the manufactured product.

Post Marketing Requirements. Once your Class II device is on the market, post-market safety surveillance should be implemented. The 510(k) approval may specify that the Class II device can only be sold as a prescription device under the direction of a physician. If not, you can sell your device as an over-the-counter device. Record keeping should include any correspondence with the FDA, clinical study data, shipping records, and a complaint file (if any complaints have been received) and the design record, including modifications.

Once a PMA order has been issued for a Class III device, the FDA will require more extensive controls beyond the general control procedures for Class I and Class II devices. For example, Class III devices will involve frequent and thorough site inspections. In fact, the FDA will need to inspect your factory and production in operation line before you can market your device, even with a PMA order in place.

You may be asked to restrict the sale or distribution of your device. You will need to evaluate continuously the safety, effectiveness, and reliability of your Class III device and to report your findings periodically to the FDA. Warnings and precautions must be prominently labeled. Device-tracking and various other precautions and restrictions that are contained in the PMA order must be implemented. In addition, the law requires that manufacturers report to the FDA any death or serious injury that the device may have caused and any malfunction that would likely cause death or injury.

If failure of the Class III device would have serious health consequences or if its implantation in the human body is for more than one year, or if the life-sustaining device is outside of the body of the device user, the FDA may ask you to carry out a post market surveillance study. If so, you, the manufacturer must submit a plan for the study within 30 days of receiving the PMA order from the FDA. The FDA will have 60 days in which to evaluate the study plan and order that the study be conducted.

8.8. Digital Health

The broad scope of digital health includes categories such as mobile health (mHealth), health information technology, wearable devices, telehealth and telemedicine and personalized medicine.

FDA expects that providers and other stakeholders are using digital health in their efforts to:

- reduce inefficiencies,
- improve access,
- reduce costs,
- increase quality, and,
- make medicine more personalized for patients.

Second, FDA sees that patients and consumers are using digital health to better manage and track their health and wellness-related activities. Equally important, FDA envisions that the use of technologies, such as smart phones, social networks and internet applications, is not only changing the way we communicate, but is also providing innovative ways for us to monitor our health and well-being and giving us greater access to information. Together the advancements in digital health are connecting people, information and technology to improve health care and health outcomes.

Many medical devices now have the ability to connect to and communicate with other devices or systems. Devices that are already FDA approved or cleared are being updated to add digital features. New types of devices that already have these capabilities are being explored. FDA's CDRH has being working to develop a guidance document that describes practical approaches to balance benefits and risks. For example, The FDA already announced that it does not intend to enforce compliance with the regulatory controls that apply to medical device data systems, medical image storage devices and medical image communications devices. The rationale is that these devices pose low risk to patients, but contribute importantly to advancing digital health.

CDRH has established the Digital Health Program, which seeks to better protect and promote public health and provide continued regulatory clarity. If you are developing a mobile health app that collects, creates or shares consumer information, you can learn from the Federal Trade Commission's website which laws of the FDA, Federal Trade Commission (FTC) or Office of Civil Rights (OCR) apply to your device of digital health.

8.9. New Drug Development on Premarket Approval Application

"From Fish to Pharmacies, The Story of a Drug's Development" is employed by the FDA to illustrate how a drug developer/sponsor can work with the FDA to bring a new drug to market.

Osteoporosis, the wasting away of bone mass, affects as many as two million Americans. The salmon produces a hormone called calcitonin. Taken by osteoporosis patients, salmon calcitonin, which is 30 times more potent than that secreted by the human thyroid gland, inhibits the activity of osteoclasts and enables bone to retain more mass. In the first segment of preclinical research, the drug developer/sponsor worked on the synthesis and purification of calcitonin. With the approval of an Institutional Review Board, the developer/sponsor carried out animal testing with an injectable form of synthesized calcitonin.

After obtaining promising data from animal studies, the developer/ sponsor submitted an Investigational New Drug (IND) application to the Center for Drug Evaluation and Research (CDER). The developer/sponsor started the second segment of clinical studies 30 days following submission as the CDER did not find any problem with the IND application in this particular case.

The following three phases of clinical studies were carried out during the second segment of research:

Phase 1. A clinical trial, involving twenty to eighty healthy subjects conducted to investigate some basic actions, dose ranges and side effects of the drug.

Phase 2. Early clinical studies, involving several hundred people to determine the effectiveness of the drug, short term side effects and risks associated with the drug.

Phase 3. Expanded controlled and uncontrolled trials, including several hundred to several thousand persons to gather additional information about the drug's effectiveness and safety in order to evaluate the drug's overall benefits-to-risk relationship.

With all supporting evidence necessary to meet FDA's requirement for marketing approval, the developer/sponsor submitted a New Drug Application (NDA) for review by the CDER's Advisory Committee. In this third review segment, the developer/sponsor may be invited to answer any questions that arise in the review. Finally, the FDA commissioner will announce the approval decision. Throughout this lengthy process, the FDA encouraged sponsor/FDA meetings.

In conjunction with this formal process, the FDA has the following provisions to make new drugs available to the public:

- Accelerated review of an NDA for drugs that have the potential to provide significant benefits for unsatisfied medical needs,
- The consideration of new, promising drugs as Treatment Investigational New Drugs so that drugs in the development process can be made available for use by desperately ill patients, and
- Conditional approval of an NDA, but a requirement that the sponsor satisfies the full set of regulations for approval after the drug is on the market.

In addition, there are post approval and post marketing requirements that drug developers and manufacturers must follow rigorously. You can find more information on topics such as patent protection of your drug and special offers for small businesses on the FDA website.

Chapter 9

The FDA 510(k) Submission

9.1. Overview

Section 510(k) of the Food, Drug and Cosmetic Act requires device manufacturers to notify the FDA of their intention to market a medical device at least 90 days in advance. This is known as a Premarket Notification (PMN) or 510(k) submission. The number of 510(k) submissions that are received may be as high as 1,200 to 1,500 per year. As many as 2,000 to 2,500 new devices enter the market each year. Many are competing products that offer incremental improvements over existing products.

To assist your preparation of the submission, we discuss issues in this chapter that are associated with the following four questions:

- Do you need to file a 510(k) submission?
- What needs to be done before the submission?
- What is included in a 510(k) submission?
- What does the FDA do following your submission?

9.2. The 510(k) Submission

When you introduce a Class I or II device to the U.S. market for the first time or propose a different intended use for your device or modify a device that you already market, you should consider the submission of the 510(k) notification to FDA. Once you have submitted this, you may receive a letter from the FDA within 90 days that contains the following:

"*We have reviewed your Section 510(k) premarket notification of intent to market the device referenced above and have determined the device is substantially equivalent (for the indications for use stated in the enclosure) to legally marketed predicate devices marketed in interstate commerce prior to May 28, 1976, the enactment date of the Medical Device amendments, or to devices that have been reclassified in accordance with the provisions of the Federal Food, Drug and Cosmetic Act.*"

This letter is enclosed with your 510(k) summary statement that bears the 510(k) number that is assigned to the device and the FDA official seal. Now, you can proceed to market your device.

There are four exceptions to this 510(k) process:

1. If you file a 510(k) and do not hear from the FDA within 90 days, you can market your product.
2. If you deem that your device is substantially equivalent to a Class I or Class II device with exemptions as stipulated by Title 21 CFR, you are not required to file a 510(k) application. It is important to document that your decision to not file a 510(k) was made in good faith and to keep a file of that documentation so that it can be made available to the FDA for inspection. Even with such precautions, you wish to avoid accumulating a very large inventory to meet market demand in case an FDA inspector informs you that you need to have 510(k) approval because the intended uses of your device differ from the uses of currently marketed devices that you have identified as being substantially equivalent.
3. If you import or distribute an approved device, you need not apply for a 510(k). If you re-label a device under your company's brand that had previously received 510(k) approval, you may need to apply for a Special 510(k), which is described later.
4. If you are not marketing the device, but using it in a clinical trial, you need an IDE (not a 510(k)) before you can proceed with clinical trial.

Two alternative approaches with shorter review times were introduced with the enactment of the FDA Modernization Act:

1. ***The Special 510(k).*** Application can be used by medical device manufacturers if the proposed device is a modification of an FDA-granted device and the modifications do not affect the device's intended use or alter the fundamental scientific technology. The Office of Device Evaluation's 30-day processing time for Special 510(k) Applications provides an incentive to submit such an application.
2. ***An Abbreviated 510(k).*** Application also can be used by manufacturers who have been able to find FDA-generated guidance

documents or information about special controls and consensus standards for their devices. The FDA has published a number of device-specific guidance documents to communicate its regulatory and scientific expectations to industry. Information is also available from the FDA about the concept of special controls to ensure the safety and effectiveness of Class II devices and about the consensus standards that are applicable to these devices. This second alternative to the Traditional 510(k) Application is easier for the manufacturer to prepare and for the FDA to review. The result is a more expeditious evaluation of the 510(k) submission and decision.

If you change the device that could significantly affect the safety or effectiveness of the device, or a major change in the intended use of the device (2), you must submit a 510(k) notification for the new device.

9.3. Preparing for the 510(k) Submission

Suppose that you decide that your device qualifies for a 510(k) submission. By this time, you should have the device in a form that is close to the final one that you will market. The first thing to do is to file at the FDA website and pay the user fee to obtain the User Fee Cover Sheet. If your company is a small business entity, you can submit that information to the FDA to obtain a small business identification number. Inputting this number into the user fee process will enable the site to reduce your user fee by half. The following items must be prepared next for the submission.

Identification of Predicate Device(s). Go to the FDA website to find predicate devices that are substantially equivalent to your device. A device can be deemed substantially equivalent to a predicate device even if it differs in technological characteristics, as long as it has the same intended uses, does not raise new questions of safety and effectiveness and is at least as safe and effective as the predicate device. Review the 510(k) summary statement of the predicate devices on the website and select the one that offers the greatest likelihood that you will be able to demonstrate that your device is substantially equivalent.

If your device has the functions of two different medical devices, they can be considered to be the predicate devices for your 510(k) submission.

Using commercial channels, obtain a sample of the predicate device in its original container and its instruction manual. Some of these materials will be used in your write-up and enclosed in your submission.

Biocompatibility. If your device will come into contact with patients or users, you will need to do biocompatibility tests (1). The FDA has issued Blue Book Memorandum #G95-1, entitled "Use of International Standard ISO 10993, 'Biological Evaluation of Medical Devices'—Part 1: Evaluation and Testing" for all premarket approval and 510(k) submissions. Device manufacturers in the United States should use this memorandum to determine which kinds of biological effects are of concern for materials that you use in your device. Such biological effects include closed patch sensitization, skin irritation, cytotoxicity, hemo-compatibility, and reproductive or developmental changes.

Software, Electric Safety and Electromagnetic Compatibility. For the 510(k) submission, you will need to know and address any FDA concerns about your software and device (4). If the software device provides diagnostic information that directly affects a decision about a treatment or therapy that could result in serious injury or death, the FDA will be very concerned. If a malfunction of the software device can lead to an erroneous diagnosis or a delay in delivery of appropriate medical care, the FDA will be moderately concerned. If these and many more similar scenarios do not apply to your device or software, the level of FDA's concern will be minor.

Similar logic applies to the subject of electric safety and your device's electromagnetic compatibility.

Sterilization and Shelf Life. It is desirable to determine if your device needs to be sterilized or what its shelf life will be, based on the materials used in the device and the device's frequency of use.

The Performance Testing-Bench, Animal and Clinical. The results presented here serve as a basis for showing the substantial equivalence of your device. The FDA recommends that the submission include the following information:

- List the specific bench tests conducted
- Describe each test protocol
    - Objective of the test
    - Test articles used in the test
    - Test method and procedures

  - Specific parameters measured
  - Pre-defined acceptance or pass/fail criteria.
- Summarize the results
- Describe your analysis
- Discuss your conclusions

Similar descriptions will be provided in the submission, if animal and clinical performance of your device were tested.

9.4. Writing the 510(k) Submission

Guidance for Industry and FDA Staff (3) and Device Advice (www.fda.gov/cdrh/devadvice) are written by the FDA to help you prepare the submission. The sections in a 510(k) notification are:

Section Description

0. TABLE OF CONTENTS
1. USER FEE COVER SHEET (FORM 3601)
2. CDRH PREMARKET SUBMISSION COVER SHEET
3. 510(k) COVER LETTER
4. INDICATIONS FOR USE STATEMENT
5. 510(k) SUMMARY
6. TRUTHFUL AND ACCURACY STATEMENT
7. CLASS III SUMMARY & CERTIFICATION
8. FINANCIAL CERTIFICATION/DISCLOSURE
9. DECLARATIONS OF CONFORMITY AND SUMMARY REPORTS
10. EXECUTIVE SUMMARY
11. DEVICE DESCRIPTION
12. SUBSTANTIAL EQUIVALENCE DISCUSSION
13. PROPOSED LABELING
14. STERILIZATION & SHELF LIFE
15. BIOCOMPATIBILITY
16. SOFTWARE
17. ELECTROMAGNETIC COMPATIBILITY & ELECTRICAL SAFETY
18. PERFORMANCE TESTING – BENCH
19. PERFORMANCE TESTING – ANIMAL
20. PERFORMANCE TESTING – CLINICAL

The Table of Contents is placed at the beginning of the hard copy. It identifies which tab to choose for each of the sections of the hard copy of the 510(k) submission. When the section number is combined with the section description to form a file name, e.g., 000_Table of Contents, the foregoing list looks almost identical to the file listing of the CD or DVD that you will send to the FDA.

Sections 1, 2, 3, 6, 8 and 9 in this list primarily constitute the administrative part of the submission. Since your device is class II, indicate in Section 7 that the section is not applicable. Sections 14 to 20 cover the test and materials that you described in the previous section. A total of 30 sections are permitted. You may use the last 10 sections to include your device's instruction manual, that of the predicate device, its labeling and packaging description, and your publications or scientific papers that are relevant to the submission. In the remainder of this section, we will discuss completing Sections 4, 5, 10, 11, 12 and 13.

Section 4 covers indications for use. It should be written like those for the predicate device. Throughout your submission, the indications for use should be identical as any differences in wording may cause the FDA staff to call for clarification. This could cause a delay.

All sections, except Section 5 (510(k) Summary), are treated as confidential by the FDA. The Summary contains the name of the predicate device(s) and its 510(k) number, a description of your device, its intended use, its performance data, a device comparison table, its substantial equivalence and the conclusion.

Once your submission has been approved, the FDA will post your 510(k) Summary on its website. Your regulatory consultant may recommend that you keep Section 5 of the Summary as brief as possible because your competitors may use this document to understand what new things you have provided to the FDA.

The FDA recommends that you describe the performance specifications of your device in the section on Device Description and include a brief description of the device design requirements. You may begin by explaining the reason for submission, which is similar to the presentation of the medical problem that you intend your device to resolve. Then, give a narrative description of the device's overall function. Next, describe the performance specifications of the device's components. Use diagrams and schematics to fully describe the function of the device and its components. Include a table about the components, part numbers and a description of their material, as well as whether the components will be in contact with the patient so that the FDA can understand the biocompatibility of your device.

You may begin the presentation of the section on Substantial Equivalence by giving an overview of the scope and analysis of what you are doing and, then, the conclusions that you have formed from your analysis.

Answer the following four questions to show that your device is substantially equivalent to the predicate device(s):

- Does the new device have the same statement of intended use?
- Does the new device have the same technological characteristics? (e.g., design, materials, etc.)?
- Are the descriptive characteristics sufficiently precise to ensure equivalence?
- Does the performance data demonstrate equivalence?

Then, a table is provided to demonstrate clearly the functional similarities of the predicate device(s) and your product, as well as their differences. Finally, provide a detailed comparison in the form of a table to summarize the similarities and differences in indications for use, device description, contraindications (conditions in which one should not use the device), sterilization and uses, technology, labeling, component characteristics and materials. Use these two tables as the basis to explain further why the differences do not affect the safety and effectiveness of your device.

The section on Proposed Labeling presents the labels that will be used on the device and its package. In general, the labels may contain some of the following information:

- Company name, logo, address and website
- Device Name
- Model name
- Key words constructed from indications for use
- Contraindications
- "Use as directed by the physician in accordance with the instruction manual"
- What the package contains
- Size information
- Patent pending, if appropriate
- Rx only?

The similarities and differences between your labels and those of the predicate device should be discussed.

The FDA encourages the use of electronic copies for Pre-Market submissions. An electronic copy is an exact duplicate of a paper submission, but is in the form of a CD or DVD that must be accompanied by a hard copy of the documents that bears the signature of an officer of the company. To provide an electronic copy, it is recommended that the preparer of the 510(k) submission use software, such as Adobe Reader 9.1, that can create PDF files with optical character recognition. This will enable the FDA examiner to use the "find" function to locate specific word(s) in the file.

Writing the submission is not the same as writing a research paper. Here are a few pointers for you to consider:

- Proving substantial equivalence must be done with products that are approved under FDA authority.
- The predicate device comparison requires an actually marketed product that has been cleared under FDA 510(k) notification or a pre-amendment device. Referencing general categories of devices to serve as predicate devices is not acceptable.
- Many clinical devices have been reported in the scientific literature with good results. Because they have not been approved yet by the FDA, they cannot be used as predicate devices to prove substantial equivalence.
- Do not use medical journal articles as proof of product effectiveness and safety. Substantial equivalence is based on intended use (or indications for use) and product claims and technology that are similar to those of existing marketed products that have already been deemed to be safe and effective by the FDA.

In this section, we provide an overview of submission preparation. Please consult the guidance documents and 510(k) submissions that are available from FDA websites (3, 4) to prepare your own submission that is specific to your device.

9.5. Outcomes

Suppose that you have a 510(k) submission that, according to the FDA checklist, is complete. The FDA has up to 90 days by law in which to reply to you. Normally, it will reply sooner. How soon is soon?

In an ideal situation, the FDA will send you a letter saying that your device is substantially equivalent to an approved device. It is also possible that the FDA will exempt your device from 510(k). If you have a letter or exemption, you can legally proceed to market your device.

However, you might also receive a letter from the FDA that states that it needs more information about your device before it can give its approval. The letter will describe in some detail what the FDA wants, and will state that you need to expand your application to include this additional information and then resubmit it. Your resubmitted submission usually can be processed in less than 90 days, as the FDA has already examined your application once.

In other scenarios, the FDA may ask you to provide more clinical data to justify your claim of substantial equivalence. Since collecting more data will take time, the FDA is likely to remove your application from their docket. If so, you will need to refile your 501(k) submission after you have the requested data.

The worst scenario is one in which the FDA finds that your device is not substantially equivalent to any Class I or Class II device. If your device is not an implant, and you think that it should not be classified as a Class III device, you can petition the FDA to create a new generic type or category to reclassify your device as Class I or II in accordance with the 513(f) process which is described in the next section. If your device is reclassified by FDA, you can proceed with your 510(k) submission.

9.6. De Novo Submission

This 513(f) or de Novo submission is the right route for your medical device if you could not identify a predicate device and you believe that your new device involves low to moderate risk. In this case, you can submit a 513(f) "Request for Information" to the FDA. The 513(f) submission should outline the characteristics of your device and include a rationale for your belief that it falls into a specific class. The FDA will evaluate your information within 60 days and issue a ruling on how your device should be classified along with a suitable regulation number and product code.

If your 510(k) submission is rejected by the FDA because the predicate device you used is not considered to be substantially equivalent to your device, but you still regard your innovative device as posing low to moderate risk, you can make a de Novo submission within 30 days after the rejection of your 510(k) application.

Since 2010, the FDA has begun releasing summary documents for devices that have been classified through the de novo process. The de novo summary is intended to present an objective and balanced summary of the scientific evidences that serve as the basis for the decision to grant a de novo request. The de novo summary also serves as a resource for the types of information necessary to support substantial equivalence for device manufacturers that may wish to use the device as a predicate for future 510(k) submissions.

A review of the Decision Summary of Illumina's MiSeqDx Platform (k123989) indicates that it is reclassified as a class II device with these intended uses: "The MiSeqDx Platform is intended for targeted sequencing of human genomic DNA from peripheral whole blood samples. The MiSeqDx Platform is not intended for whole genome or de novo sequencing." Within that summary document, the FDA evaluates the analytic performance of the platform with data on

a. Accuracy: Three accuracy studies were conducted.
b. Precision/Reproducibility: Two reproducibility studies were conducted.
c. Linearity: Not applicable.
d. Carryover: A study was conducted to evaluate the potential for inter-run and intra-run sample carryover. Intra-run sample carryover tested the system in the most challenging scenario for sample carryover within a single sequencing run.

In another summary document on the Genetic Health Risk Test (DEN 160026) of 23 and Me on 11 diseases, data on the performance studies done on each disease are presented. It also contains a section on benefit/risk analysis that concluded "Given the device's indications for use, required general controls and special controls established for this device, the probable benefits would outweigh the probable risks."

References

1. Wallin R. F. and P. J. Upman, MD&DI, January 1998. www.devicelink.com/mddi/archive/98/01/023.html
2. www.fda.gov/MedicalDevices/DeviceRegulationandGuidance/GuidanceDocuments/ucm080235.htm
3. www.fda.gov/MedicalDevices/DeviceRegulationandGuidance/GuidanceDocuments/ucm084365.htm
4. www.fda.gov/MedicalDevices/DeviceRegulationandGuidance/GuidanceDocuments/ucm089543.htm

Chapter 10

Obtaining Help and Support

According to the website of the Small Business Administration (SBA), the 23 million small business entities in USA employ almost half of the private work force, produce half of the private sector's output, and account for 97% of all US exports (7). Almost 40% of jobs in high technology sectors are in small businesses. Small businesses contribute significantly to the US economy in both innovation and job creation.

In 2008, there were approximately 6.1 million businesses. About 627,000 of them were new, 596,000 have closed and 43,000 have filed for bankruptcy. It has been suggested that one out of seven new businesses grow and earn some profit. In recognition of the risks and importance of small businesses, the government and private sectors provide many forms of support to help them to succeed. In this chapter, we will discuss:

- Support from the SBA for small businesses
- Government grants to sponsor the research and development of small businesses
- Community support for small high tech businesses.

10.1. SBA Support for Small Businesses

Small Business Administration. The SBA was created in 1953 as an independent agency of the federal government to aid, counsel, assist and protect the interests of small business concerns, to preserve free competitive enterprise and to maintain and strengthen the overall economy of our nation. As explained in its website, the SBA has four programs to assist your business development. They are:

- ***SCORE*** (Service Corps of Retired Executives). This is an SBA resource partner that is dedicated to entrepreneur education and the formation, growth and success of small businesses nationwide.
- ***SBDC*** (Small Business Development Center). This provides

management assistance to current and prospective small business owners.

- ***WBC*** (Women's Business Centers). These form a national network of nearly 100 educational centers that are designed to assist women start and grow small businesses.
- ***SBA District Offices***. The SBA operates full service district offices in every state of the nation.

The assistance of SBA and local government is oriented toward the service industry, some retail, and a few manufacturing and construction industries. Even though the needs of ventures in biomedical industry may differ from those of the mainstream industries that receive SBA assistance, support from these four programs and their partners will be useful for the growth of biomedical ventures.

Business Development. Your business development may help you to find a city in which to locate your company, to develop a viable business plan, and to obtain financial assistance. The local SBA office and government will be happy to help you to develop and grow your company.

To help you in your decision to build a factory in a particular locality, the economic development office of the local government, the SBDC, and the Chamber of Commerce can provide information about the labor force that is available for employment, the names of comparable companies that are situated in that locality and the availability of local resources such as space, roads and tax incentives. The possibility to collaborate with various laboratories of the university at that locality will be another factor that will help you decide where to locate your business.

What should you do to start or grow the business? How do you prepare a strong business plan, resolve business problems, acquire management skills and find resources to support the business operation? These are the problems that the SBDC and its local partners can help you to resolve. A number of attorneys, certified public accountants (CPAs), bankers, and small business people serve in the SBDC to assist its operation and course teaching.

If you wish to buy a business, sell a business or start a business, the SBDC will advise you how to go about it. It can advise you of the local rules and regulations for your business, which permits you will need and how to structure your business. With regard to the last-mentioned subject, the SBDC can help you to set up your business as a sole proprietor, partnership, limited liability company (LLC) or corporation. In the area of sales and marketing, it can advise you of the ways to carry out an initial market analysis and put your marketing plan into action. If you intend to market your service locally, the SBDC will help you to identify your competition, regardless of your market, and the demographics of your market – number and size of households, median income, etc. It can help you set up your accounts or books and also advise you on the long-range outlook for your market. For information about running a small business, there are a number of training courses or workshops that the SBDC and its partners offer. The impact of the SBDC on the Commonwealth of Virginia is mentioned next.

Virginia, which has a population of 7.6 million persons, has 29 local SBDCs. According to the websites of these SBDCs, more than 4,500 business owners and managers received one-on-one counseling in 2006 and more than 9,200 attended local SBDC's training programs. SBDC professionals assist with business planning, marketing, financial analysis, access to capital, business start-up and other specialized services that are requested. Many companies have given an account of the impact that the SBDC has had on their operations and attributed much of their success to their relationship with the SBDC. In 2006, loans of more than $73 million were obtained, total capital of $109 million was formed, 4,170 jobs were created or retained and sales increases of $99 million were reported.

The SBA has a web site that spells out in detail how to prepare a business plan. The SBDC firmly believes in the need for a business plan. First, a business plan describes what you do when you start a business. You need to keep working on the business plan and applying it to your business rather than flying the business by the seat of your pants. You will constantly updating it - at least annually - as suggested by the SBDC. Very often, an entrepreneur has a great idea, is technically knowledgeable and can usually find someone to put money into it. However, when it comes to managing the business and really making it successful, the entrepreneur's business skills are not always that great. SBDC views the business plan as something that compensates for many weaknesses. Once you put your business development ideas down onto paper and have experts, like bankers or investors, look at your business plan, they can tell you if you have a good grasp of the business and where you will need help with the operation. As described later, the SBDC will help you to obtain financial assistance from various SBA programs.

SCORE is a sister organization of SBDC. Retired CEOs, company presidents, and businessmen serve as volunteers to staff it. There are more than 10,500 SCORE volunteers in 374 chapters that operate in more than 800 locations to assist small businesses with business counseling and training. SCORE also operates an active online counseling initiative. You can ask their counsellors to review your business plan and introduce you to people who can assist your business development.

Financial Assistance. If you want to establish an office to practice your profession or build a manufacturing facility that can generate income within months, the topics that are covered in this subsection will help you to obtain financing to develop your business.

Loan assistance is a big part of SBDC's operation. Generally, SBA is not in the business of making loans, but it does provide loan guarantees. You will want to use commercial bank financing to take your company to the manufacturing and marketing phase or to put you into a position to obtain VC investment in the future. At this time, you don't have much collateral or an extensive track record, especially if your enterprise is a start-up. The bank will likely consider your small business loan application, if you can secure an SBA guarantee. Just as your parents may co-sign a loan for you, the SBA may agree to guarantee part of a bank loan in order to help you to secure the bank loan. The SBA's loan guarantee mechanism is known as the SBA's 7(a) Standard Loan Program. An SBA pre-qualification loan and a Small Business Investment Company are other avenues that are available for small business financing.

How much money can you obtain from a bank? If you have a business plan that shows that you need $1 million to carry it out, the SBA will guarantee 75% of the loan of the required funds and you, as an owner, must come up with the remaining 25% of the required funds. In essence, the maximum loan will be $750,000 and you will have up to 25 years to repay it. For smaller amounts, the SBA may guarantee 85%, leaving you to come up with 15%. The loan can be used to pay for a building, equipment, working capital, and salaries. The funds you personally invest constitute your stake in the development of the business.

At this point, most banks will want you to submit a loan application, including a business plan. Without a plan, most banks won't even discuss a loan. The bank's greatest concern is how you will repay the loan. In the business plan, you must be able to show your sales projections, the anticipated expenses, what will remain after you pay all expenses, how much will be available to repay the bank and also to pay yourself a living wage or salary. The bankers look at the 4Cs of you and your business development. They look at your character (Have you ever been in jail?), your credit (What is your credit record and have you ever been bankrupt?), your capital (How much capital can you put into it?) and your collateral (What collateral can you provide?). If you have a patented invention, completed its R&D and built a pre-commercialization model that will serve as your collateral. With this collateral and a good business plan, the SBA will enthusiastically provide a loan guarantee and the bank will not turn you down for lack of collateral such as real estate or securities. What the bankers want is a good business plan, a good product, good character, a good credit record, and some investment in the venture by you. The SBDA will help you to develop the loan strategy, package the business plan and present it to the bank. If you like, its representative can accompany you to the bank to help to explain the proposal.

The SBA pre-qualification program is designed to help small businesses in certain locations of the country and also those that are owned by minorities or women. In the case of a husband and wife who are starting a business, a 51% ownership by the wife will qualify the venture for this particular program. Also, if you're a veteran or have a rural business, you can qualify. The other requirements for loan guarantee are similar to those of 7(a) Loan Program, except that the loan amount limit is about $250,000.

For-profit Small Business Investment Companies are licensed and regulated by the SBA. They provide management assistance to start-up and businesses in their early stages. The financial assistance is a combination of debt and equity investment. If you approach these companies, they will work to find investors who can supply investment capital to your company. They will also provide some debt financing. The investors will own part of the company. The rest of the financing will be provided as a bank loan with an SBA loan guaranteed.

10.2. Government Support for Technology Development and Commercialization

Small businesses contribute much more to the US economy and society than can be calculated from the spending and profit that they generate. These businesses are more innovative than larger companies and more able to respond to changing consumer demands. Small businesses produce more patents per employee than do large firms and small businesses patents are twice as likely than patents held by larger firms to be among the 1% of patents that are most cited (i.e., most significant). These comments were made by Derek Liebert, an Adjunct Professor of Government at Georgetown University (3).

The development of biotechnology products requires years to accomplish. Some products require a large capital investment to carry out the necessary research and development and to commercialize them. As Hector Barreto, SBA Administrator, stated (2), "But small businesses cannot do it alone. The US government plays a crucial role in creating an environment in which entrepreneurs can flourish....Government investment in small businesses benefits not only the businesses themselves, but also our national economy and our society at large."

Many state governments have programs or agencies that focus on entrepreneurs who have a record of innovation. These federal and state programs are reviewed in this section to provide you with an opportunity to decide whether you would like to apply for federal and state funding to carry out research, development and commercialization.

Federal Grants. In 2001, nearly $1.5 billion in federal funding was awarded to small businesses to develop technology concepts or products. By 2007, the Small Business Innovation Research (SBIR) programs of 11 federal agencies had awarded $12 billion to various small businesses. The SBIR program is a "set-aside program "in which the agencies allocate 2.5% of their research funding for domestic small businesses to engage in R&D that shows potential for commercialization. In 1992, the Congress authorized the establishment of the Small Business Technology Transfer (STTR) program to support small businesses that are working to commercialize technologies that were developed in universities or non-profit research institutions.

The National Institute of Health of the Department of Human Health and Service is one of the federal agencies that is most likely to fund the R&D of your medical device or drug. NIH regularly offers SBIR Conferences throughout the nation. The technical and administrative staff of NIH man these conferences at one-on-one tables to discuss NIH's interest in SBIR, project planning, proposal development, cost accounting, intellectual property, commercialization, and forming teams with the medical industry and universities. State agencies, universities, and private organizations partner with the NIH to offer workshops to help small businesses to understand the issues that are important in SBIR programs, to make more competitive grant proposals, and to develop strong commercialization plans.

In many cases, you can strengthen your research program and enhance your chances of receiving SBIR funding, if you partner with well-established investigators or institutions to carry out the R&D. For example, you may subcontract part of the SBIR grant to a certain investigator of a university to carry out clinical trials or basic animal studies of your product, because he can provide facilities and expertise that your small business does not have. Conversely, the university faculty may collaborate with you to apply funds that are received from the NIH to examine the impact of your product on certain diseases. A contractual agreement that specifies the rights and responsibilities of each party to the collaborative project is essential to ensure that both your company and the university benefit. This agreement is especially important if you plan to apply for STTR funding.

To learn more about the application process for SBIR funding, please refer to Chapter 11.

Supports from State Governments. In recognition of the tremendous potential in permitting the private sector to commercialize the research findings of universities, Virginia established the Center of Innovative Technology in 1984. The State hopes that job opportunities, tax revenues and other economic benefits will be gained by helping universities to commercialize their intellectual properties through the CIT (3). Many states have similar agencies, which are identified at the www.gnet.com website. We will use CIT to illustrate more specifically what these agencies do.

In order to help universities to carry out cutting edge research in all areas of technology, CIT helps to create potential applications of university research in business and industry, to accelerate the transfer of that knowledge to the marketplace, and to promote the growth of entrepreneurial firms and start-ups within the state.

CIT's mission has evolved into the following: "to accelerate the development of the next generation of technology and the building of technology companies through its research investment and entrepreneurship programs." CIT's goals are to expand Virginia's technologies into world-class research hubs and to enable Virginia to become a development leader in entrepreneurial technology ventures.

CIT and its partners offer SBIR conferences, TechStart Bootcamps and Proposal Assistance Awards to help you to win SBIR/STTR grants. CIT has been instrumental in helping Virginia companies to win R&D dollars for high-risk technologies from the Advanced Technology Program of the US Department of Commerce. The Federal and State Technology Partnership (FAST) Program that is operated by CIT can assist companies to solicit government funding, especially from the federal government. Virginia universities also receive funds from the CIT to operate Technology Innovation Centers to facilitate collaboration among researchers in universities and companies.

The CIT has nine regional offices in Virginia. Because the CIT is committed to the support of companies in Virginia, the regional directors are knowledgeable in all aspects of business development, including market analysis, intellectual property protection and the commercialization of products. One of the regional directors has a Ph.D. in microbiology and really enjoys the variety of the work for the CIT, start-up companies, and university research laboratories. These regional directors offer one-on-one consulting, technical data and information about your industry, access to expertise in universities and federal laboratories and referrals to a statewide network of partners who know your industry.

The CIT office in Charlottesville serves the Central Virginia Region and has close relations with the University of Virginia. There are many start-up activities in Charlottesville that receive support and advice from the CIT. Its staff is focused on advanced materials, advanced manufacturing, biotech, electronics, information technology and the telecom sector. To monitor trends in these industries and funding opportunities, steering committees have been established and staffed with experts in the advanced materials, electronics and biotech areas.

The CIT maintains a database of all patents that have been issued in Virginia. This helps Virginia universities to market their patents and you, as an entrepreneur, to find someone in the Virginia system who has something beneficial to contribute to your company and whose location is not too distant. CIT also helps with patent licensing. If you need something concerning the patents that are owned by a university, the CIT will be of great assistance. The university's patent office also will help young start-ups to get their businesses off the ground.

10.3. Community Support

Charlottesville Venture Group. Many localities have private organizations that promote business and technology development. For example, Charlottesville and Central Virginia have the Charlottesville Venture Group (CVG). This is an organization that is staffed by volunteers and committed to smarter ventures and smarter investments through education (6). It has a quarterly publication, "Capital Talk," that covers various topics, such as what to exclude from in a business plan. One of CVG's activities is the Business Spotlight, which conducts a business plan competition. This is intended to give you an opportunity to present your plan to an audience of venture capitalists, entrepreneurs, and other interested members of the Charlottesville and surrounding business community. CVG is partnering and collaborating with various organizations in Charlottesville to bring their wealth of expertise in a wide range of business topics to the members and to build entrepreneurial activity, which is essential to business growth.

CONNECT. This is a technology and business accelerator that is dedicated to the rapid and sustained growth of high technology and science companies that have a vested interest in the San Diego region (5). Serving as a proven neutral broker, CONNECT directly addresses the critical business needs of companies, ranging from start-ups to large public corporations, as well as universities and entrepreneurs. CONNECT is supported by more than 2,000 volunteers, many of whom are seasoned entrepreneurs, CEOs of successful companies or attorneys of various specializations. Below are some of the programs operated by CONNECT.

- **Access to Capital**
  - **Springboard and Springboard Spotlight**. Mentoring to companies in all stages of development for three to five months by Entrepreneurs in Residence. Quarterly Spotlights profile the graduates for investors.
  - **Venture Roundtable**. Connecting venture capitalists to early stage venture opportunities in technology, diagnostics and devices, life sciences, therapeutics and clean technology.
  - **Tech Coast Angels**. A network of private investors in early-stage Southern California companies.

- **Entrepreneur Education**
  - **Entrepreneurs FrameWorks Workshops**
  - **CEO Strategy Forum**
  - **MIT® Enterprise Forum**
- **Networking and Recognition**

  Programs and opportunities that are specifically designed to bring the community together to create a dialogue, recognize innovations and identify successful individuals and companies.

So successful is CONNECT that San Diego is on its way to becoming the premier region in the country for innovations.

Tech Coast Angels. This is the largest angel network in the United States (9). It has 280+ members who have invested more than $100 million in over 130 start-ups in Southern California. The San Diego Chapter has an annual Quick Pitch program that enables companies to present their business plans and network with the venture community. As a result of the efforts of Tech Coast Angels, the sponsored start-ups have gone on to raise a total of $1 billion in venture funding.

10.4. Genetic Sequencing by Synthesis Technology

On the morning of February 28, 1953, James D. Watson and Francis H. C. Crick of the University of Cambridge determined that the structure of DNA was a double-helix polymer, or a spiral of two DNA strands, each of which contains a long chain of monomer nucleotides, wound around each other. According to their findings, DNA replicates itself by separating into individual strands, each of which becomes the template for a new double helix. They solved a fundamental mystery of science, namely how it is possible for genetic instructions to be held inside organisms and passed from generation to generation. This discovery began the race to sequence the DNA, to study genetics and molecular biology and to develop gene and cell therapy.

It wasn't until the 1970s that two groups developed distinct sequencing chemistries that held considerable promise (1). At Harvard, Cambridge, USA, Allan Maxam and Walter Gilbert developed an approach that exploited the selective chemical reactivity of reagents towards the nucleobases to induce sequence-specific cleavage patterns from which the sequence could be derived after separating the cleaved DNA fragments, based on their size, by high resolution gel electrophoresis. In Cambridge, UK, Frederick Sanger and his co-workers developed a very different approach. It involved the synthesis of DNA in the presence of nucleotides that terminate synthesis.

Sanger sequencing proved to be the practical method of choice. It subsequently opened up the field of genetics and molecular biology. The best-known, and arguably most important, application of Sanger sequencing was the completion of the first human genome in 2004 by The Human Genome Project. A copy of the human genome comprises more than three billion bases of DNA. It was decoded over a period of several years using hundreds of capillary sequencers in a collaborative effort that involved laboratories worldwide, at an overall cost of $2.7 billion.

This important milestone provided the opportunity for humanity to begin to understand our genome on an unprecedented level. It has also stimulated the consideration of alternative approaches to nucleic acid sequencing that have gone beyond what was possible using Sanger's approach.

In the summer of 1997, a new idea dawned on S. Balasubramanian and D. Klenerman of the University of Cambridge/ It was the idea of using clonal arrays and massively parallel sequencing of short reads with the use of solid phase sequencing by reversible terminators (subsequently referred to as sequencing by synthesis technology) as a DNA sequencing technology. They subsequently approached Abingworth Management, a venture capital firm and obtained initial seed funding to form Solexa. In 2001, the team's research progress attracted $14 million in Series A funding. This enabled Solexa to build its management team and to acquire molecular clustering technology from Manteia. The amplification of single DNA molecules into clusters enhanced the fidelity and accuracy of gene calling, while reducing the cost of the system optics by generation of a stronger signal.

Illumina went on to purchase Solexa in an all-stock deal worth $600 million in 2007. In a press release at that time, the two companies stated that the market opportunity in genetic sequencing was estimated at $1 billion. Illumina has since built upon sequencing by synthesis technology, expanded the parallel sequencing to millions of units, and improved the original technology for faster sequencing at lower cost.

By 2016, Illumina had reduced the cost of determining one genome sequence to about $1,000 and the time to five to ten days. According to the OmicsMaps website, Illumina has 1,145 machines in place worldwide. This compares to 836 for all other vendors combined! Since Illumina machines generally produce the highest yields, they are responsible for an even larger proportion of the total bases of DNA sequence information assayed worldwide.

This technology development described the situation for a case that has received help and support from venture capitalists, governments and universities. However, the winners are still decided by the market.

10.5. 3-D Printings and Organs on a Chip

3-D printing for manufacturing is one of the 21st century's most exciting technological breakthroughs. This technology is most widely used in the automotive industry where it helps produce prototypes of new cars or car parts. The next largest market is the aerospace industry, where manufacturers are using the technology to make lighter versions of complex parts for airplanes.

The technology dates back about 30 years. What makes it exciting, however, is that costs of printers have come down in price drastically and the "printing" materials have greatly expanded. For world healthcare, the market for 3D printers was estimated to be worth $579 million in 2014 and is estimated to reach $2.4 billion by 2020 with a compounded annual growth rate of 26%. If pieces of skin, liver or kidney using live cells can be printed, analysts estimate the medical market could reach a value of $6 billion or more within 10 years.

In 2015, Joseph DeSimone and his colleagues published a paper on a process that can continuously form a part within minutes (4), instead of hours with the layer-by-layer method of 3D printing. In this continuous liquid interface production (CLIP), the liquid resin at the bottom of the part is activated by ultraviolet light to solidify while the part is being elevated (Fig. 10.1). This exciting technological development will not only further reduce the manufacturing cost and also generates finer and more complex products, such as the micropaddles that are shown in Fig. 10.2 of page 163.

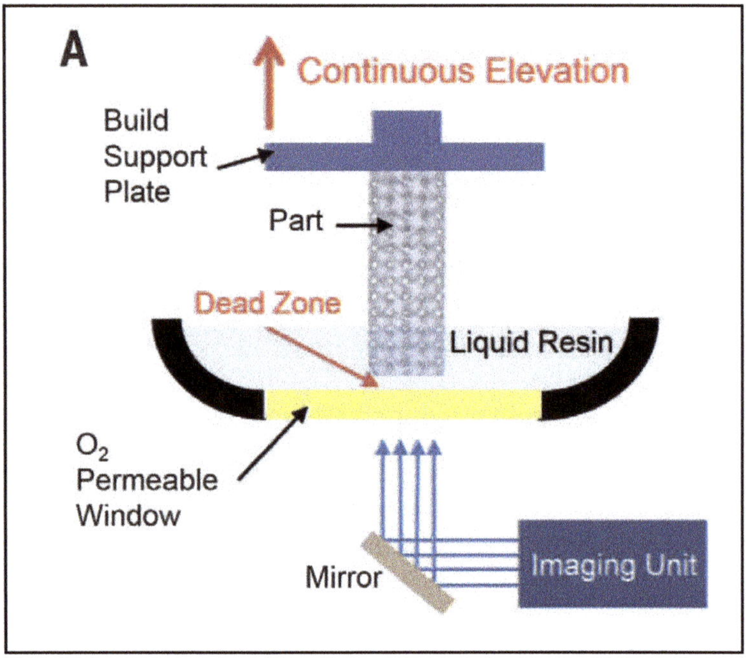

Fig. 10.1 The system of Continuous Liquid Interface Production showing the solidification of liquid resin activated by light generated by the imaging unit. (Courtesy of *Science* and DeSimone, 4).

Desimone took a leave of absence from North Carolina State University to found Carbon3D to further develop and market this CLIP technology. He was awarded the Heinz Award in 2017. "*Dr. DeSimone's achievements as a polymer scientist and entrepreneur leading to singular breakthroughs in areas such as 3-D printing, nanomedicine and green chemistry are many, and the positive effects on how we live, create, work and treat our planet are only just beginning to be seen,*" said Teresa Heinz, Chairman of the Heinz Family Foundation. "*We honor him with the Heinz Award in Technology, the Economy and Employment not only for these accomplishments, but also for his ability to work across the traditional boundaries of scientific discipline, and for taking knowledge gained out of the laboratory and into the places where it can have a positive impact.*"

Fig. 10.2. The details of micropaddles with stems of 50 μm diameter that are produced by CLIP. (Courtesy of Science and DeSimone, 4).

"*3D manufacturing will enable a world of new applications in virtually every manufacturing industry, from automotive and aerospace, to electrical equipment, to consumer-product industries. The technology will finally realize its full potential as an essential manufacturing tool,*" said DeSimone. "*Another area of exceptional promise is medicine. The technology could be used to produce a host of medical devices on-demand and personalized for the individual patient, such as new forms of drug and vaccine delivery vehicles. Surgical implants — such as stents — could also be quickly produced to match the dimensions of a patient's individual anatomy.*"

More information on CLIP can be found on the website of Carbon3D.

In a *Trends in Biotechnology* special issue on biofabrication that was published in August 2017, researchers reviewed and considered the progress that has been made in 3-D bioprinting and what might be possible in the decades, or years, ahead. Here is a condensed version that was released by Cell Press (8).

1. Made-to-Order Organs-on-a-Chip. "Organs-on-a-chip," 3-D, micro-engineered systems that mimic the structure and function of human tissue, are a strong contender in the race to deliver inexpensive and efficient personalized medicine. Lung, gut, and pancreatic tissue have already been grown from human stem cells on chips, which allows researchers to study physiological differences in these cells between patients, as well as to screen for drugs.

2. Skin Manufacturing Printed skin made from cells set down on a collagen gel have shown the presence of intercellular connections and biologically normal cell markers 10 days after cultivation. In another study, researchers have been able to grow blood vessels in a sheet of cells. Skin bioprinting is closer to reality than one would think. However, researchers are only beginning to consider the designs necessary to help patients, especially those who have burns or chronic wounds.

3. Facial Reconstruction Although bone, cartilage, skin, muscle, blood vessels and nerves have been printed in the laboratory, constructing more complex designs that can be implanted in patients is still in development. Craniofacial reconstruction, which would benefit people with cancer or who have experienced facial injuries, seems to be an obvious candidate to pursue because of the amount of work already done on these cell types. In the short term, 3-D printed scaffolds could be used to improve spot defects in the jaw or other areas of the face.

4. Multi-Organ Drug Screening. 3-D bioprinting is demonstrating that precise models can improve the way in which we evaluate new drugs, such as by generating "organoids" that are made of multiple cell types, as well as a tumor model with engineered blood vessels. Although such approaches could make it possible to quickly monitor drug interactions in real time in multiple organs, much more iteration (e.g., adding blood vessels, connecting organ models) will be needed to realize this vision.

5. Plug-in Blood Vessels. Efforts to create 3-D blood vessel networks within bioengineered tissues--which would be necessary to ensure tissue survival after implantation and an accurate replication of human anatomy--have focused on stacking 2-D layers of cells or bioprinting 3-D networks, which permits high levels of spatial control. One challenge is to create tissues with blood vessel networks that could connect directly to a patient's arteries or veins.

In conjunction with development #5 a group at Northwestern University in Chicago in 2016 printed working prosthetic ovaries for mice. The recipients were able to conceive and give birth with the aid of these artificial organs. No one is yet talking of printing gonads for people. However, blood vessels are a different matter. Sichuan Revotek, a biotechnology company based in Chengdu, China, has successfully implanted a printed section of artery into a monkey. This is the first step in trials of a technique that is intended for use in humans.

The trends in developments #1 and #4 are further highlighted by the following two press releases. First, the National Center for Advancing Translational Sciences (NCATS) of NIH awarded $17 million in 2014 to support the next phase of its Tissue Chip for Drug Screening program. These grants are made to improve ways of predicting drug safety and effectiveness. Researchers in 11 institutions will collaborate for a period of three years to refine existing 3-D human tissue chips and combine them to create an integrated system that can mimic the complex functions of the human body. These programs in the 11 institutions will receive additional support over the remaining two years, if funds are available.

Because these tissue chip systems will closely mimic human function, scientists can probe the tissue chips in ways that they aren't able to do in people, and the knowledge gained may provide critical clues to disease progression and insights into the development of potential therapeutics.

During the next phase, researchers also will increase the use of induced pluripotent stem cell (iPSC) technology as a renewable human cell source for their systems. iPSCs are derived from adult cells that can be reprogrammed into embryonic-like cells that can then be turned into other tissues. A goal of the Tissue Chip for Drug Screening program is to increase efforts to create a single iPSC line that can differentiate, and mature into, all major organ systems in the human body. At present, researchers can use iPS cells to generate different tissues that are used in some of the tissue chips. However, no single iPS cell line can produce yet all of the needed major organ tissues.

The website of the Wyss Institute for Biologically Inspired Engineering at Harvard University describes the function of an airway on a chip for chronic obstructive pulmonary disease (COPD) and asthma (10). COPD affects 23 million Americans and the yearly cost of treating asthma in the US is approximately $56 billion. A research team at the Wyss Institute leveraged its organ-on-a-chip technology to develop a model of the human small airway in which lung inflammatory diseases, which are the third leading cause of mortality worldwide, and asthma can be studied outside of the human body. As reported in Nature Methods, the platform enables researchers to gain new insights into the disease mechanisms, identify novel biomarkers and test new drug candidates.

COPD and asthma are inflammatory reactions in the lung that can be dramatically exacerbated by viral and bacterial infections, as well as smoking. It is known that many of the associated disease processes occur in the conducting airways of the lung that shuttles air to and from the alveoli or air sacs. However, much less is known about how inflammation induces distinct pathological processes, such as the recruitment of circulating white blood cells and the buildup of mucus that compromises the lungs of these patients, or how clinical exacerbations are triggered.

"*Inspired by our past work using the organ-on-a-chip approach to model the lung alveolus, we created a new microfluidic model of the lung small airway that recapitulates critical features of asthma and COPD with unprecedented fidelity and detail. Now with this microengineered human lung small airway, we can study lung inflammatory diseases over several weeks in chips lined by cells from both normal donors and diseased patients to gain better insights into disease mechanisms, as well as screen for new therapeutics*," said Donald Ingber, M.D., Ph.D., the senior author in this work who is leading a multidisciplinary team of Wyss scientists, which has been at the forefront of organ-on-chip technology.

"*This novel ability to build small airway chips with cells from individual patients with diseases like COPD positions us and others now to investigate the effects of genetic variability, specific immune cell populations, pharmaceutical candidates and even pandemic viruses in an entirely new and more personalized way; one that will hopefully increase the likelihood of success of future therapeutics*," said Ingber.

References

1. Balasubramanian S., Sequencing nucleic acids: from chemistry to medicine, *Chem. Commun. (Camb).*, 47: 7281-7286 (2011).
2. Barreto, H.V., *eJournal USA — Economic Perspectives*, 11:1 (2006).
3. Leebaert, D., *eJournal USA — Economic Perspectives*, 11:3 (2006).
4. Tumbleston JR, …., DeSimone JM, Continuous liquid interface production of 3D objects, *Science*, 2015, 347:1349-1352www.cit.org
5. www.connect.org
6. www.cville-venture.org
7. www.sba.gov
8. www.orthospinenews.com/2016/08/12/five-ways-bioengineers-want-to-use-3-d-printing/
9. www.techcoastangels.com
10. wyss.harvard.edu/modeling-copd-and-asthma/

Chapter 11

SBIR Grants for Product Development

11.1. Vision of Small Business Innovation Research (SBIR) Program

In recognition of the risk of failure of high-tech, small businesses, the Small Business Innovation Development Act of 1982 sets aside 2.5% of federal agencys' extramural research funding to support the R&D of domestic small business entities. In accordance with this legislation, NIH awarded $566 million of SBIR grants and contracts and $73 million of STTR (Small Business Technology Transfer) grants to small businesses in 2008 (1). This investment has enabled small businesses to accomplish a great deal in taking new medical products to commercialization, in order to improve the delivery of healthcare and the health of people, and to contribute to the economy of the nation during the last twenty five years.

The mission of NIH is to improve human health. Your medical device or pharmaceutical product development fits into this mission. NIH will be interested in evaluating your device or drug for SBIR grant funding consideration.

Since tax dollars are used for this investment, NIH has established an extensive and thorough review process to evaluate the innovations and strengths of SBIR/STTR proposals that are submitted to NIH and to prioritize the funding of these proposals.

Obtaining a VC investment for R&D for pre-commercialization of a device or drug is very challenging. One uncertainty for VC is that your business model has not been shown yet to be feasible or the device or drug to be marketable. SBIR grants prepare you to advance from the product's feasibility and R&D phases to the manufacturing and marketing phase. Thus the SBIR grants would greatly strengthen your position to obtain VC investment to commercialize your products.

11.2. NIH SBIR Funding

NIH has two types of SBIR funding. Almost 90% of the funding is used to support SBIR grants. For SBIR grants, your company decides on the R&D that you wish to undertake and the product that you want to develop and commercialize. The project depends completely on your own initiative. If you feel that the project is important to medicine, you submit a SBIR grant application to NIH for R&D funding of a pre-commercialization device or a drug. The second type of funding is by an SBIR contract. NIH needs certain R&D done or products to be developed, but has no in-house people to do it. NIH will offer to fund small business entities to carry out the specific R&D under a contract. A much smaller number of grants are also awarded in the form of STTRs, These are somewhat similar in form to an SBIR grant, but with greater emphasis on technology transfer.

In addition to the requirement that the company must have no more than 500 employees, the business must be owned by US citizens or permanent residents who own at least 51% of the enterprise and have control of the research facilities. The 51% ownership requirement prevents the subsidiary of another company receiving an SBIR grant. The aspect concerning control of research facilities also requires the applicant to have a place to carry out the proposed R&D. Thus, the company can either rent space for the R&D activity or own space for it so that no one, except the applicant, can use that facility.

In general, your company activities can be divided into three phases:

- Phase I: Assessing the feasibility of making your product and evaluating its technical merit
- Phase II: Conducting research and developing your product to its pre-production model
- Phase III: Commercializing your product.

The government will support the first two phases. Funding for Phase III is expected to be provided by non-SBIR/STTR sources.

Suppose that you propose the development of a medical device to achieve a certain sensitivity and that a device of such sensitivity is not available in the market. The scope of this project will fit well with a Phase I SBIR grant from NIH to undertake R&D to demonstrate its feasibility (i.e., that you can produce your conceptual model of the device). Of course, you must demonstrate that the project has technical merit. This means that it must provide important technology to solve clinical problems. For a Phase I grant, the R&D guideline requires you to have a budget of no more than $100,000 for a period of six months.

The Phase II grant is intended to enable you to continue the R&D that was initiated in Phase I. In other words, the preliminary results obtained during Phase I will be the basis of your Phase II SBIR grant application submission. The guidelines used by NIH for Phase II grants limit the total budgeted costs to $750,000 and the period to two years. When you have sufficient results to demonstrate the feasibility and technical merits of your device, you may consider submitting a Fast Track Award to obtain both Phase I and Phase II funds from NIH. Before you embark on this track, you may wish to ask NIH about others persons' success in obtaining Fast Track Awards and the percentage of funding set aside for this track.

R&D in genetic engineering or medical development can be expensive. The R&D may require the use of expensive reagents or the development of a new genetic line of animals. Thus, NIH will allow an exception if you can justify your larger proposed budget. You must consult the NIH staff if you have a budget that exceeds the guidelines of $100,000 for Phase I and $750,000 for Phase II.

The budget contains two types of costs - direct costs and indirect costs. Direct costs include the salaries of the principal investigator (PI) and researchers who do the R&D, the equipment and supplies required to perform the tasks, attendance at national meetings and any R&D work that are to be subcontracted. The rental of space, bookkeeping charges, and the salary of the PI to manage non-R&D supporting activities are indirect costs. Indirect costs are usually calculated as a percentage of the direct costs. However, for the percentage that is applicable to the operation of your small business, you should consult with NIH. You can choose to set a small percentage of the total costs as profit for your company.

The funds received from SBIR grants differ from a bank loan, as there is no need to repay the government after your company becomes profitable. Consult the government about expenditures that you cannot charge to the grant. For example, NIH thinks that the patent application is a by-product of your R&D. Thus, you cannot use SBIR grants to pay filing fees for patent applications. However, the government normally grants your company the ownership of the patent of the invention. This gives the company the right to earn royalties.

11.3. NIH SBIR Grant Review Process

NIH and the Department of Defense are the federal government's two main agencies for funding basic research at universities and the R&D of small businesses. NIH has a peer review process that ranks the priority of grants for funding. Respectable researchers and university faculty members (the peers) serve on study sections to evaluate grant applications and establish funding priorities. The grants manager of NIH relies completely on the recommendation of the NIH study sections. Because other agencies have specific mission to fulfill, their managers have considerable leeway in deciding whether to fund your grant.

To ensure that you receive an exceptional review, NIH invests time, manpower and money to make the review system work fairly and effectively. If your grant application receives no funding, you will receive a constructive review and valuable feedback to help you to revise your research plan for another SBIR application.

The research plan that you submit for an NIH SBIR grant is reviewed for five aspects:

1. Significance of the research
2. Viability of the proposed approach
3. Innovativeness of your project
4. Training and experience of the investigators
5. Suitability and adequacy of the research facility.

The first three aspects deal with the technical merit of the proposed research and the last two deal with the ability of the small business to carry out the R&D. The reviewers will look at these five aspects and give you a score, which will decide whether your grant will be funded.

After you submit an SBIR grant to NIH, NIH will assign it to a study section for review. If you wish, you can recommend a particular study section to review your grant. In total, the NIH study sections review 5,000 to 6,000 new applications each year. Some sections include bioengineering or medical imaging in their section name and biomedical engineers serve as reviewers for many sections. Each section has 20 to 30 members. They meet three times a year in Washington, D. C. to review and score about 60 to 80 grants during each two-day meeting.

As an example of the reviewers' qualifications, one SBIR grant study section has three members who have a dual degree of PhD/MD, 12 members with a PhD and 10 who hold an MD. Three of these reviewers come from industry and the rest are employed by universities or research institutes. The executive director of the study section appoints two to three reviewers to serve as the primary reviewers of your grant application. In the review meeting, these primary reviewers summarize for the review panel what you propose to do and how you will carry it out. They will discuss the merit of your proposals and give the panel a feeling of how high they would score your grant proposal. The next task is to triage the proposals into three categories. These are: insufficient merit for funding, sufficient merit for consideration and good proposals that require detailed consideration. After further discussion, each member submits a score (1 to 5, with 1 designating best) for the grant proposal. The score is given in increments of 0.1. Then, the executive director of the study section adds up the scores and multiplies the average by 100 to yield a final score.

The greater the score, the lower it is in priority for funding. The final score and the reviewers' comments and evaluation are conveyed to you. The best possible score is 100. This would imply that you can be absolutely certain that you will obtain the NIH grant. If the score is 200, your research proposal is probably on the borderline for funding. The total fund allocated to each R&D topic decides the last grant that will be funded on that topic. If you receive a score of 300 or more, your proposal is unlikely to receive any funding from NIH.

Whether the PI under an SBIR grant should have a Ph.D. or an M.D. is an issue that has been raised. NIH's official statement is that this is not necessary. On the other hand, NIH does require you to demonstrate that the PI has sufficient expertise to oversee the research projects scientifically and technically. Describe your education, scientific and engineering training, and research experience in the grant application to show that you are qualified to be the principal investigator. To assess your qualification, the review panel critically evaluates your research publications in refereed journals.

11.4. Research Plan of SBIR Grant Application

The main body of the research plan should contain these five sections:

1. Specific Goals
2. Background and Significance
3. Preliminary results (not needed for Phase I grant applications)
4. Research Design and Methods
5. Product Development or Commercialization Plan (not needed for Phase I grant applications).

In the first Section on specific goals, you spell out what you want to do. Then in the second section you describe the background and significance of what you want to do. In it you also address why this SBIR grant is important for you and why it is important for NIH to support your SBIR development. When you have carried out your Phase I R&D, you describe the preliminary results, supporting the need for further R&D in the third section. The fourth section describes how you will do it. Finally, for Phase II grant application, you have the fifth section to overview your plan to commercialize your product.

The page requirement for the three key sections of the Phase I research plan is 15 pages and 25 pages for the first four sections of a Phase II grant. It is recommended that the commercialization plan be no more than 10 pages in length. The other parts of the research plan include human subjects, vertebrate animals, literature that has been cited, contractual arrangements and consultants.

NIH prefers that you do not proceed directly to patient studies because it wants to be assured by animal studies that your invention is safe before you test it on humans. It is important that you set a few realistic goals for your Phase I R&D that you can accomplish in six months with the $100,000 budget. For example, you may set a goal of developing a prototype of a blood volume monitor. Larger or more ambitious aims, such as to develop a fully functional blood volume monitor and to test it with patients, cannot be done within the constraints of time and budget of a Phase I grant. If you exceed your budget, the NIH will not supplement your grant to pay for the difference. You can receive no cost extension of your Phase I grant for another six months. Similar budget constraints apply to the operation of Phase II grants. No cost extension can be one more year.

Suppose that you accomplish the goals of the grant during Phase I. Presenting this evidence in the preliminary study section in the Phase II grant application will help to convince the reviewers that you can do what you propose in that application.

To prepare a good application, read the *Advice on NIH SBIR & STTR Grant Applications* by Gregory Milman, National Institute of Allergy and Infectious Diseases (gmilman@niaid.nih.gov). This presentation of 88 slides provides a great deal of excellent advice that may help you to succeed with your NIH SBIR or STTR application (2). In addition, you should obtain a funded SBIR grant application to learn how to do it successfully. NIAID has an outstanding Phase II grant application that is available for your examination (3). Equally important is to have a seasoned researcher to review your grant application and to incorporate his suggestions to improve your grant application well before its deadline for submission.

To facilitate the transition of SBIR or STTR Phase II projects to the commercialization stage, Phase II Bridge Award will be available to Phase II grantee if the applicant can secure independent third-party investor funds that equal or exceed the total amount of the NIH grant.

11.5. Timelines for SBIR Grant Funding

It normally takes about one year from submission of your grant application to receive the funding from NIH, if approved, to begin the Phase I research. If you completed the Phase I research satisfactorily and then submitted a Phase II SBIR grant application promptly, it may take slightly less than one year to receive Phase II funding. In essence, you need to be prepared well in advance of that time with whatever funding may be available to pick up the slack between Phase I and Phase II of NIH SBIR funding.

Suppose that your company has two products for development and that both can be done, overlapping somewhat in time. You can submit a proposal for a grant for one product, and then submit another grant for the second product in the following year. If you are able to receive grant award notices in accordance with the timelines suggested above, you will have continuous funding to develop your two products for five years. Again, this continuous funding arrangement can be disrupted if the funding for one application is not granted due to a low score.

References

1. www.nih.gov
2. www.niaid.nih.gov/ncn/sbir/advice/advice.pdf
3. www.niaid.nih.gov/NCN/sbir/app/default.htm

Chapter 12

Writing and Presenting Your Business Plan

12.1. Overview

A business plan is multi-purpose. First, writing it forces the executives to organize their thoughts, plans and dream for the company into a business plan that can serve as a blueprint or roadmap to build the venture into a successful and profitable business. The plan provides a conceptual picture that answers the following questions:

- What do you plan to do?
- Why do you want to do it?
- How you are going to achieve success?

The plan provides a map that lays out the goals, timelines and actions for everyone in the company to follow and the company's financial goals. The plan identifies the employees' responsibilities and describes the benefits that the company's product provides to patients. Thus, it serves as internal guidelines for the employees in carrying out their work and moving the company ahead in its mission to aid healthcare. The plan motivates potential employees to join the company, which offers more powerful diagnoses or treatments for patients. Externally, the business plan can be utilized as a means to stimulate the support of existing investors and attract new capital from outside sources, such as venture capitalist firms (VCs).

Business models describe how a company develops and markets its product for revenues and profits. In the biotechnology sector, there are three primary business models (6). They are: 1) The Fully Integrated Biotech Model, which is committed to developing the pharmaceutical product, to conduct clinical trials, to develop its value and, finally, to market it, 2) The Technology Platform Model, which is to develop a platform technology and then sell the technology to enable a buyer to discover its own novel drugs or drug targets and 3) The Royalty Model, which licenses the product or technology to a large pharmaceutical company in exchange for a royalty. In your market research, you will learn what business model is used by other companies that have products similar to yours and use the market research to select the right business model for your company.

In this book, we address the development of the company in three phases. They are Assessing the Venture, Launching the Venture and Building-Up the Enterprise. The product development, actions to take, and financials aspects required for each phase differ. As you begin the development of your company in the Assessing Phase, it is important that executives and employees know what needs to be done. Thus, it is desirable to prepare the business plan as a roadmap for company development. When you launch the company and have many actions to carry out, you need a roadmap that has been rewritten as an Action Plan. Both the Roadmap and the Action Plan have the structure of a business plan, but are more in line with that phase of development. As you move into to the Building-up Phase, you will need capital to carry out your plan to develop manufacturing, marketing and sales. Thus, you should write a Business Plan that has an emphasis on raising investment capital.

Writing the business plan for your venture, therefore, is an evolving process. For example, as you modify the Roadmap to become the Action Plan, you will need to update the presentation details that you already have, add the new action items to undertake during the Launching Phase, and, then, revise your future projections of company developments. Before preparing any of these plans, it is important to know what your company has and, also, has achieved. This is addressed in the next section. The subsequent three sections cover guidelines and suggestions for preparation of the Roadmap, Action Plan and Business Plan.

Many companies can get the company off the ground with funds from founders, friends or families. As capital needs increase, you are bound to look for funds from VCs. Aside from the obvious need to access capital with which to fund the growth of your company, you may be interested in leveraging the VC's knowledge and relationships into a strategic partnership for the development of your company.

Successful venture capital firms are approached frequently by companies that need financing. Therefore, it is critical that your business plan defines clearly and succinctly the market opportunity that you are targeting and why and how your company is better positioned than your competitors to capitalize on it. In this chapter, we explain the process of writing the business plan and presenting it to your internal team and the investment community.

Your business plans include proprietary information about your company. When you show a plan to potential employees, you can ask them to sign a non-disclosure and confidentiality agreement. Signing such an agreement is not the custom for VCs. Thus, it is advisable to do due diligence on potential investors to determine how ethical they are before you send them your business plan with a confidentiality reminder.

In writing your business plan, it is recommended that you read the book *The Business of Healthcare Innovation* by Lawton R Burns (2). This book provides a detailed overview of the pharmaceutical, biotechnology, genomics/proteomics, medical device, and information technology sectors. It analyzes the market structures of these sectors as well as the business models and corporate strategies of firms operating within them.

12.2. Company Status before Preparing the Roadmap

Must Have. Once upon a time, while several biomedical engineers were having a friendly discussion of a hot research topic in a private setting, they came up with an innovative concept for a medical device that may benefit patients. After several brainstorming sessions, they had:

1. Gained a commitment from several persons to work together in an entrepreneur team to build up a company. One of them had the training, expertise and experience that make him an ideal choice to serve as the Chief Technology Officer (CTO) and the PI of a Phase I SBIR grant application to demonstrate the feasibility of building the device by state-of-the-art technology.
2. Completed a conceptual description of the innovative device and what it may do for patients.
3. Completed a literature survey showing the importance of the medical problem that the innovative device is intended to resolve.
4. Concluded from the number of patients affected by the medical problem that the potential market was large.
5. Discussed the device and the company with several angel investors. This led to one investor's promise to invest $100,000

for seed money to get the company going, if a satisfactory business plan for the new start-up is submitted.

6. Interviewed and consulted several experts who are knowledgeable about the particular medical problem. They were enthusiastic about the development of the new product and its potential to solve the medical problem. The experts agreed with the entrepreneur team's belief that the functionality of the product was superior to those devices that competitors marketed. These experts indicated their willingness to serve as consultants to the company.
7. A vision and a company mission statement .

If you are in a comparable situation and can point to achievements like those mentioned above, you can go ahead and prepare the roadmap.

The Vision and Mission of Your Company. The vision is the company's guiding statement and response to the following questions:

- Why do you want to develop that medical device?
- Why do you want to market it?
- How does your device benefit patients?

In effect, you use the vision statement to tell the VCs and anyone else why you're doing what you do.

A very good example is the Microsoft Corporation's vision statement (4). The vision of Bill Gates and his partners, who founded the company, is to put a personal computer on every desk and in every home. They thought that this vision would serve to meet the needs of the coming Technology Century by helping to improve productivity, efficiency, the search for information, and so on. This vision addresses the problem that they saw at that time - the lack of operating systems that could be used easily by common people. Following the tremendous growth of Microsoft, its vision now empowers people through great software anywhere at any time and on any device. Such a mission statement tells people that, with the products of Microsoft, they can accomplish much more than they've done in the past.

The benefits of the vision (3) are that:

- A vision forms the basis of extraordinary human effort,
- A vision provides a context for strategic and tactical decisions,
- A shared vision creates cohesion, teamwork, and a community,

- A vision lays the groundwork for a company to progress beyond dependence on a few key individuals.

The mission statement, the focus of your Company's vision, should address the following question. What does your Company want to achieve? If you want to put a computer on every desk, you should be talking about hardware, software to manage the operating systems and various applications for office and home use. Because Microsoft had received a contract from IBM to develop DOS for personal computers, Bill Gates and his associates decided to concentrate their mission statement on the development and sale of software to enable Microsoft to become the largest provider of PC operating systems, software and all programming tools required for various applications, such as office operations and hospital information management, etc.

The mission statement of Medtronic (1) is:

- *To contribute to human welfare by application of biomedical engineering in the research, design, manufacture, and sale of instruments or appliances that alleviate pain, restore health, and extend life.*
- *To direct our growth in the areas of biomedical engineering where we display maximum strength and ability; to gather people and facilities that tend to augment these areas; to continuously build on these areas through education and knowledge assimilation; to avoid participation in areas where we cannot make unique and worthy contributions.*
- *To strive without reserve for the greatest possible reliability and quality in our products; to be the unsurpassed standard of comparison and to be recognized as a company of dedication, honesty, integrity, and service.*
- *To make a fair profit on current operations to meet our obligations, sustain our growth, and reach our goals.*
- *To recognize the personal worth of employees by providing an employment framework that allows personal satisfaction in work accomplished, security, advancement opportunity, and means to share in the company's success.*
- *To maintain good citizenship as a company.*

The first statement is really Medtronic's vision. Here is the story told by Earl Bakken on the adoption of this statement by Medtronic and the impact it had on Medtronic as the company reached the age of 25 (1):

"Medtronic's financial data in the early 1960s revealed the bleak reality (i.e., near bankruptcy). *The exhilarating growth that followed the development of, first the external pacemaker and then its implantable successor was hugely expensive for a small, undercapitalized, relative inexperienced company like ours.*

It didn't take us long, then, to decide what we wanted to concentrate on as a company. We would dedicate ourselves in the development, manufacture and sale of devices that restored people to meaningful lives. We would focus on implantable therapeutic technologies – as opposed to diagnostic or other laboratory products – and build the company around those technologies. Our written mission statement would reflect that decision.

My role, I decided, was to encourage our engineers in their activity, remind them of the corporate mission, help them see the big picture, and make sure they didn't limit their imaginations or stifle their creative urges.

In 1974, Medtronic celebrated its 25th anniversary. For those of us present at or near the company's creation back in 1949, such a milestone scarcely seemed possible.

For a company that has established itself as a leader in the field of biomedical engineering, is the world's largest manufacturer of pacemakers, and will soon gross $100 million annually, Medtronic certainly had a modest beginning some 25 years ago."

12.3. Roadmap for the Assessment Phase

With the company scenarios that were described in the previous section, you can incorporate the answers to the following two questions in a business plan or roadmap for the company. The roadmap is written to convey the message that the company has a good plan for pursuit of three short-term goals and one-long term goal:

- To recruit qualified executives and employees who will help move the company forward
- To demonstrate the feasibility of manufacturing the medical device
- To prepare the company to attract funding with which to move the company to the next level of development

- To become a market leader with the sales of effective medical devices.

First Question: What Do You Have at present? The scenarios that are described in the previous section are used to answer this question by indicating that your company has the rudiments of a promising entrepreneur team, a conceptual model of the medical device, and the endorsement of an angel investor and several medical experts on the development of the company.

Many other scenarios can result in similar answer to this question. In one situation, a successful CEO, who had just sold his/her medical device company, was visiting a biomedical engineering professor to learn more about the research that was being done in the professor's laboratory. As a post-doc described his/her research project, it dawned on the ex-CEO that this project could evolve into the development of a new medical device. As a result, the CEO and the post-doc at that laboratory quickly formed a new company and soon progressed to the stage of preparing a roadmap for the company.

In a networking party that a local entrepreneur chapter and a research university organized, several invitees had the opportunity to discuss their careers and work. Pretty soon, they discovered a common interest in the use of a new technology to deal with a certain medical problem. After these individuals had met for a few brainstorming sessions, they appointed a CEO and a CTO and established a company with funds from their families or their own savings.

Second Question: What Do You Want to Accomplish in the Coming Six-months or Year? The answer to this question can be explained by a plan that describes the execution of the following four tasks for the scenarios described in the ***Must Have*** segment of Section 12.2:

- Recruit a CEO and establish the organizational structure and responsibility chart
- Undertake further study of the market to identify the technology, competition, and barrier to entry
- Submit a Phase I SBIR grant to conduct the feasibility study
- Develop the research plan to carry out the feasibility study.

Research Plan of Feasibility. This third task will be addressed by what you prepare the research plan of your Phase I SBIR grant application. The 15-page research plan details the objective of your development project, the design and construction of your medical device, the parameters that you are measuring, the sensitivity that is required, and the preliminary data that you have. All these are used by grant reviewers to learn if what you proposed to do is feasible and your grant can achieve a score that is fundable.

General Guidelines on Writing a Business Plan. A business plan consists of an executive summary, five sections as the main body of the plan, and supporting documents. At this early stage, it is recommended that the executive summary be limited to one quarter to one half of a page in length and that the main body consist of no more than 10 pages. Supporting documents are not included in the page count. In presenting the process to write the three plans, *the part of the guidelines that apply to the writings of all three plans is italicized.*

The Roadmap for the Assessment Phase is prepared by combining your answers to the two questions posed previously in this section with the responses to the general guidelines below into one to convey the message that is underlined in the first paragraph of this section.

Executive Summary. This summary highlights two messages: your technology is the best in resolving the medical problem faced by many patients and your team can build the company to become the market leader. *Construct an elevator speech of 30 seconds from the summary to convey the two messages to potential investors and employees.*

1. The Company. *This section may cover the following topics:*

- *The vision and mission of the Company*
- *The organization of the Company*
- *The entrepreneurial team*

• The background, past experience and capabilities of the CTO, who has the qualifications necessary to obtain funding for an NIH Phase I SBIR Grant and to get the technology developed

- *Current ownership*
- *Investment received and present financial status*
- *The product that is being developed and the medical problem to be resolved.*
- *Organizational development*
- *Product development*
- *Market development*

Since the developments to pursue in the last three subjects during the Assessing Phase are covered in the next three sections, any description of them in the Company section would be short, as the latter will concentrate primarily on the long-term development of the company. Use the material that you collected for the list of achievements in the previous section for a discussion of the company vision and mission, organization, entrepreneur team, technology, and development plans in the Company section.

2. Technology Development. In general, *this section is used to show that your product has the following four characteristics:*

- *Its technology is really the best available for solving the medical problem*
- *The technology provides a competitive advantage (5)*
- *Your product can achieve a high penetration rate and can gain a large market share with a minimal entry barrier*
- *Your product can generate profits for your company.*

In the Roadmap, you use the materials that were collected from interviews and literature to support your assertion that your conceptual model can soon be shown to possess the four characteristics above.

Suppose that it is your plan to secure an SBIR Phase I grant to fund your feasibility study, which will show that you can build the conceptual model into a functional device. You will use one page in this section to describe the research that you will undertake with the Phase I grant.

Conclude this section by outlining the technology that the company will develop in coming years.

3. The Management and Organization. *This section describes the establishment of a management team under an organization structure that will efficiently handle the business side of product development, employee management, marketing and sales.* Use the materials that you collected in the previous section to write this section and the next. Some organizational structures for the company are suggested in Chapter 6.

4. The Manufacturing Plan. *This section describes the actions to take to have the product manufactured at an affordable cost.* In writing this part of the Roadmap, you describe the research to conduct so that the investors will know how you will overcome the obstacles in manufacturing your products.

5. The Marketing Strategy. *This strategy describes how the company will engage competitors, identify and segment customers and market your product to the actual and changing market environment.* It also describes how you deal with the five forces of Michael Porter (7): a) threat of new entry, b) threat of substitution, c) buyer power, d) supplier power and e) competitive rivalry.

6. Financial Aspects. *This section presents the financial projections to show that the company is able to use its resources and capital in this phase to undertake the R&D and facility expansion that are described in the business plan,* i.e., the Roadmap. Do a cash flow analysis to show that you give close attention to financial matters and will not be caught in a situation with insufficient funds in your bank account because of inadequate planning. Use your knowledge of the market and your product to estimate the market share that you can probably obtain in five years. Next, prepare a projection of revenue and expenses to show when your company will become profitable.

7. Supporting Documents. *Enclose at the end such documents as the executives' CVs, scientific publications concerning the product, an abstract of the patent, the product's instruction manual, and the FDA letter for receiving the 510(k) application, if available.*

12.4. Action Plan for the Launching Phase

Action Phase. Once the company has recruited a competent CEO and obtained a satisfactory result from the feasibility study, the CEO and CTO can proceed to convert the Roadmap into an Action Plan. The message of the Action Plan is that you will soon be able to show that your product has a sustainable competitive advantage and will generate sales that will create a profit for the company. In writing the plan, you should emphasize the following points:

- The research done in the Assessing Phase and for the Phase I SBIR grant puts the Company into the position of being able to do the work on infrastructure, manufacturing, quality and regulatory control that is necessary to move the company to the manufacturing and marketing of the medical device.
- The updated market survey and evaluation of the competitive advantages of the medical device further supports the assertion

that the company can become the market leader in the field that the medical device serves.
- The tasks to be done will enable the question of whether a prototype of your device can be constructed at a reasonable cost. This question is known as the cost question that the Launching Phase will address.
- The submission of an SBIR Phase II grant for R&D.
- The sales of medical device will increase the value of the company and provide a good return of investment (ROI) for the investors.

In addition, you indicate in the plan how you will prepare the following:

- A prototypic device that can be mass produced and demonstrated by means of a Phase II SBIR grant.
- A detailed cost analysis to manufacture the device
- A demonstration of its diagnostic or treatment power by collaborative research with physicians and biomedical engineers
- A detailed documentation of the cost effectiveness of the device and the potential for insurance reimbursement
- A projection of the revenue generated by sales of your device and associated disposable products
- A non-provisional patent application for the device
- A 510(k) submission for Premarket Notification of your Class II medical device

The Action Plan covers a period of two to three years with a funding at the level of $500,000 to $1.5 million to accomplish the tasks that are listed above. (See also Table 6.3 for comparison.) A good Action Plan can be used to apply for funding from angel investors and venture capital firms to speed the completion of these tasks and to begin the work of the Building-up Phase sooner.

To rewrite the Roadmap into an Action Plan, you need to strengthen the parts in the Roadmap that describe what you have accomplished during the Assessment Phase. Then, you use Chapters 5 to 11 to lay out a management strategy that can convince internal personnel and investors that the management team members are goal achievers. With improvement in these two parts, you next revise the part that describes what you plan to do in the future. Limit yourself to half a page to a page for the executive summary and 15 pages for the total Action Plan.

More Elaboration. To complete the Action Plan, four more items should be discussed further. The first is the submission of a Phase II SBIR grant. This will be dealt with next. Although your class II device may not require clinical trials, it is desirable that you look for others to collaborate with you in clinical studies of your device as a second action item to carry out in this phase. The clinical studies will determine whether your device will benefit the treatment or diagnosis of patients. The results should be published in a scientific journal. Thirdly, provide more detailed milestones (including dates) to mark your company's progress during the next five years, especially for the following:

- Completion of the R&D for a prototype
- FDA approval and start of production
- The first sales
- Reaching the breakeven point
- Achieving the full market share that you had projected.

Finally, provide a description of the exit strategy and show, by the financial projections, how good a return the investors will receive on their investment and what the employees will receive for their work.

Phase II SBIR Grant. One task for the Launching Phase is the submission of a Phase II SBIR grant. Achievement of its overall objective would show that you have a prototype that is patented, approved by the FDA for marketing, shown to have clinical benefits to patients, outperformed your competitors, can be manufactured at a reasonable cost and fulfill the requirements for insurance reimbursement.

12.5. Business Plan for the Building-Up Phase

With the completion of the action items that are proposed for the Launching Phase, you move to the Building-up Phase. This involves constructing manufacturing capability, testing the facilities for quality and regulatory assurance and forming the marketing team to sell your medical device and its disposable products. The message to be delivered about this phase by the Business Plan is that your management team will move the company forward into the manufacturing and marketing of medical devices that are highly beneficial to patients and generate a good ROI for investors.

What you have done in the Assessing and Launching Phase will enable you to write a strong Business Plan and probably cause your Company to become known by the community as a bright, young company. With the Business Plan, a prototype and publications that show its medical benefit, you have a better chance of raising the needed capital to build up your Company on terms that are acceptable to both you and the investors.

In addition to improving the Action Plan and including the building plan for manufacturing and marketing, you can add more supporting evidence to strengthen your Executive Summary and Financials. The Business Plan should be limited to 20 pages, including an Executive Summary of one or two pages.

Executive Summary. This summary is written with the goal of stimulating the interest of investors to invest in your company. You need your summary to stand out so that the investors will be convinced that you have a promising company and they will receive a satisfactory return on their investment.

The Company. By including new information obtained in the Launching Phase, this section of the Action Plan and the following three are revised to become the corresponding section of the Business Plan.

The Technology. In this phase, you may wish to conduct a user satisfaction survey in order to obtain information to use in refining your device design. Since it will be necessary to manufacture many devices to carry out the survey, you will gain some information about manufacturing costs that will enable you to confirm that you can mass produce your device at an affordable cost.

The Management. In addition to the revisions, describe your plans for hiring supervisors and workers for production, quality control, regulatory affairs, servicing, marketing and sales.

The Manufacturing and Marketing Plan. Use the materials covered in Chapter 15 and 16 to prepare this plan. Include in the plan the establishment of a service and quality control unit. Calculate from your research of marketing, pricing and insurance reimbursement, a more realistic projection of the market share that your product can gain after it has been introduced. Use this projection when you do the financial calculations of the next subsection.

Financials and Exit Strategy.

Investor Funding. Stipulate here what funding you will be looking for from investors and how it will be used for infrastructure, equipment acquisition and personnel development, etc.

Assumptions for Financial Analysis. Here is a list of values and factors that you enter into your financial software to do the projections:

- Investment side
    - Investor funding
    - Earnings before interest, taxes, depreciation and amortization (EBITDA) multiple
- Marketing and revenue
    - Percentage penetration for estimation of the full market share
    - Number of years to achieve the full market share
    - Pricing of the device and consumables
- Expenses in each of the following four major categories:
    - Cost of goods sold
    - Expenses of sales, marketing, servicing and regulatory and quality control
    - R&D expenses, including clinical trials
    - Administrative expenses, such as salaries, directors' fees, travel, office supplies, equipment, licensing fee, shipping, storage fees, rents, maintenance and utilities, marketing and advertising.

Financial Computations. The computations will be done for a period of five years in the expectation that the exit strategy described next may be implemented at the end of the fourth or fifth year. Here is a list of projections that investors may wish to examine:

- The profit and loss statement
- The expense statement showing major expense categories
- Cash flow
- Number of employees
- Use of investor' funds before breakeven is achieved
- Stock price
- Net present value
- Rate of return on investment

A 5-year financial analysis of a monitor company is given in Section 13.5 on Financial Projections.

Sensitivity Analysis. First, the projections can serve as a means to check the validity of your business model. If the projections are bleak, you may need to identify the remedial or contingency actions that you should take to correct the deficiencies in the business model and/or assumptions. You should do a sensitivity analysis to examine how a change in the assumptions can affect the financial projections and the impact of that change on the company's profitability and the return on investment. For example, assume that the 2008 financial crisis (if it ever happens again!) produces a 20% decrease in sales. The sensitivity analysis with this parameter change incorporated can project how much longer it will take to reach a breakeven and what the return on investment will be at the end of the fifth year.

Exit Strategy. The investors can receive a return of their investment by means of:

- buy-outs by employees
- a merger or acquisition by another firm
- an IPO (initial public offering)

They certainly would like to know your preference in this matter.

12.6. Presentation of the Business Plan

Quick Pitch. As you progress into the Building-up Phase, you may be invited to participate in a business plan competition. San Diego Tech Coast Angels is an organization that holds an annual Quick Pitch Competition in which you can present your venture, expose your company and network with members of the audience and of a panel that includes VCs, angel investors, bankers, accountants, lawyers and executive recruiters (8).

The competition organizers may ask you to submit a brief description of your company and your management team's experience in no more than 200 or 300 words for screening. In your description, you may answer questions, such as those following:

- What is the market for your product?
- Why will the patient or physician use your device?

- What are your competitive advantages and IP protection?
- How will you make a profit?
- How successful has your team been in managing other ventures?

If the screening committee feels that your venture will be of interest to the panel, you will be invited to make a two-minute oral presentation aided by two to four slides. You may wish to expand your presentation to include the following additional points:

- Marketing and sales strategy
- Financial projections and milestones
- Funding request and its use.

With a time limit of two minutes, you will not be able to share all of your excitement about the venture with prospective investors. Concentrate on topics that will excite and interest the audience in two minutes, so that you will be invited to a more comprehensive business plan presentation or to submit your business plan for their evaluation.

The Objective of the Business Plan Presentation. You should set the objective you want to achieve as:

- To explain the company's venture as clearly as possible
- To create excitement in the audience about the company
- To secure a second meeting with the investors.

Following the presentation, you may have time for a Q&A period.

Format for Slide Preparation. You can use 15 to 20 slides to make this 15-minute presentation. Please note that each presentation should be unique with its own flow.

In preparing your slides, it is important that you not burden your audience with too much information. It is desirable that each slide contain no more than eight lines and fewer than 40 words. More importantly, each slide should contain only one or two key messages. The other points mentioned in the slide should be there primarily to support the key messages. Use graphics to better convey your message. Back-up slides could be pulled out to elaborate on points that might be raised at the Q&A session.

Template of Subjects Covered by the Slides

Slide 1. The mission of the company. It is also used to introduce the company's name and the name of the presenter and his or her position in the company.

Slide 2. Accomplishments and any awards of the key company executives who lead the entrepreneur team. List in the slide any patents granted and submitted.

Slide 3. The device and its unique features.

Slide 4. What you want. You can list two or three needs. One can be funding to raise and the other can be a Vice President of Sales, for example. Further elaboration and justification can be provided later.

Slide 5. The patients who are served by the device. Use this slide to show the number of patients who are likely to benefit from use of the device.

Slide 6. The medical problem that the device addresses. This is also known as the "pain" slide. It tells why patients and physicians will really want to use your device for treatment or diagnosis.

Slide 7. Market opportunities. Use a short number of facts or statistics to show the size of the potential market.

Slide 8. Funding that the company has raised and milestones that it has reached. This slide describes what has been done and with what funding.

Slide 9. Benefits provided to clinics or hospitals by your device. Because of these customers' financial constraints, you should show that there will be better care along with a reduction of the cost of the operation by using your device. Use this slide to explain why clinics and hospitals will want to buy your product for use by patients.

Slide 10. Competition and competitive advantages. Describe the problems in using the device marketed by your competitors.

Slide 11. Feedback from patients and physicians on the use of your device, if you have this information available.

Slide 12. Go-to-market strategy. Do you want to make direct sales to clinics and hospitals, to serve only as a wholesaler, or to function as an Original Equipment Manufacturer (OEM)?

Slide 13. The company's financial projections. You can list some of the assumptions made and use one more slide for this subject.

Slide 14. Milestones to reach. These include obtaining FDA approval, beginning the promotional campaign and making sales.

Slide 15. Summarize the strengths that are unique to your company.

Slide 16. Provide three to five points that you want the panel to remember after they have left the room.

Slide 17. Repeat two or three questions that you want the panel to focus on.

Slide 18. Exit strategy for investors. You may let it remain on the screen during the question and answer period.

Summary Remarks. Prepare a solid business plan to deliver the message that your technology is the best and that your team can build the company into the market leader. Then you will succeed in building a strong company and making good sales,

References

1. Bakken, E.E., *The Man's Full Life*, Medtronic, Inc., pp. 84-93 and p. 150 (1999).
2. Burns, L.R., *The Business of Healthcare Innovation*, Cambridge Univ. Press (2005).
3. Collins, J.C., and W.C. Lazier, *Beyond Entrepreneurship*, Prentice Hall, p. 53 (1992).
4. Gates, B., Quotes, en.thinkexist.com/search/searchquotation.asp?search=microsoft+vision&q=
5. Olsen, E., entrepreneurs.about.com/od/beyondstartup/a/whatwarrenwants.htm
6. Pfeffer, C.G., The biotechnology sector — therapeutics, in *The Business of Healthcare Innovation*, ed. LR Burns, Cambridge Univ. Press, p. 122-131, (2005).
7. Porter, M., Porter's 5 forces, www.mindtools.com/pages/article/newTMC_08.htm
8. www.techcoastangels.com

Chapter 13

Financing and Accounting

13.1. Angel Investors and Venture Capital Firms

Financing a start-up depends largely on the business, its potential and environment, and its financial needs. In today's environment, venture capital is hard to obtain. Angel investors or joint venture deals with well established companies are the best sources. Banks are not likely to provide financing that you want if you have no cash flow with which to service the loan. Because it will take considerable time to get your invention to market, it is important that you establish a second means of activities to create revenue for the company. This can be the marketing of existing products that are similar to yours and consulting services.

The importance of healthcare to people has motivated many engineers, scientists and physicians to found companies in the areas of biotechnology and medical device development for diagnoses and treatments. These companies will be poised to contribute greatly to the world's healthcare in this Medical Device and Biotech Century. Angel investors and venture capitalists will recover from the present financial crisis to invest in, and promote these entrepreneurial developments.

Angel Investors. Many angel investors are successful entrepreneurs who have been benefited from angel investments. The reason why they invest in start-ups is to receive a better return than offered by traditional investments. They are seasoned investors and know fully the risk in start-up investments. Angel investing is usually a sideline activity for these investors. They can provide good management advice that will help you run the company. It is less time consuming to obtain angel investments than to secure financing from venture capital firms. Their due diligence is less intense than that of VCs, as described in the next section. Networking, business plan competitions, and web searches are some of the ways you can use to identify angel investors who may be interested in investing in your company.

Some angel investors are known as FFF, families, friends and fools. They may be new to venture investments and may make their decision to invest largely on the basis of their trust in you. Although you may be confident that their investment in your company will grow, remember that a majority of new businesses fail during their first three or four years. Thus, it is your responsibility to make sure that these investors fully understand that they assume a risk when they invest in your company. Do not accept money if it comes from savings and pensions that the investor depends on for retirement. It is better to have a life-long friendship than to cause a financial crisis for him in his later years if your company does not succeed.

The financial transactions of start-ups involving these individuals and other angel investors are seldom publicized. As a result, information about angel investments tends to be somewhat anecdotal. The profile of these angel investors, as described below, should be considered only as guidelines when you decide whether to attempt to secure angel investments.

- Angels invest about $20 to $30 billion per year in approximately 50,000 to 60,000 companies.
- The average angel investment is about $30,000 to $40,000.
- Investors expect an annual rate of return of 25% to 30%, or to recover their entire investment three or four times in three to five years.
- Angel investors believe that one third of their investments are likely to fail.

To keep informed, angel investors normally require quarterly reports to assure themselves of your company's progress.

However, there are angel investors who want to have greater control of the company and/or to have a seat on the board of directors. This can be equivalent to acquire another partner for your entrepreneur team. Structuring the relationship between investors and the company right away is essential for the company's progress of the company.

A good source of funding to call attention to is **high net-worth individuals** who are well established in business and periodically invest. Once committed to an attractive deal, these individuals can move quickly. Typically, these individuals are less sensitive to company valuation than VCs or angel investors. The amounts that can be raised from this source typically range from $500,000 to $750,000.

A Distraction in an Investment Deal. The first venture of legendary biotech entrepreneur Snyder was the Cambridge Mailing Machine Company. It manufactured and sold machines to use to stuff envelopes at high speed. He told the following story to a class of Biomedical Engineering Entrepreneurship at the University of Virginia. The events in the story occurred in 1964:

A famous lawyer in New York found me and said, "You know, Sherry, I hear through someone at the golf club about a machine you have. It's fascinating, and I'd like to invest in it. How much money do you need?"

I said, "Well, I probably need $150,000 to get it out in the market."

He says, "Fine, fine. How much money have you got?"

I said, "Twelve-thousand dollars is all the money I've got."

He said, "I'll put in $12,000, you put in your $12,000, and we'll have 50/50 split of the company. Then I'll loan you the balance." He said, "The only condition is that I'll have a market research study made and as long as it says the market is as big as you say it is, we will go on, and the loan will become a five-year long-term note. But if it doesn't, I can call the loan and get a hold of the patents and the company."

I had confidence enough to know what this figure was, so I said, "That sounds pretty good to me on the 50/50 split." And off I went.

A few months later, the lawyer's market study was done, and actually it said the market was about twice the size of the market figures I'd given to him. All of a sudden, I get a call from his lawyer, and he said, "Jeez, I'm sorry, Sherry. We're really disappointed by this market study, and I think we're going to have to call our note."

I said, "Wait a second. The market study is twice the number of my market estimate."

He said, "I'm sorry," and they called the note.

Now, this was around October, and I had until December 31 to pay back all of the money he had put into the company, which was about $50,000, or else I had to sign the patents of the machine and the company to him. Talk about being up against the wall!

(Through another cliff-hanging experience, Snyder received an investment of $74,000 in November.)

I waited until December 30, when the New York lawyer was ready for the transfer of the company. I sent a check to him and told him what he might want to do with it, expletives deleted!"

According to Snyder, "*God takes care of young, well-meaning entrepreneurs.*" However, the lesson here is you should be cautious about entering into contracts. Pay special attention to the fine print. And, even after you do, remember that the hardest-working and most ingenious entrepreneurs sometimes still need luck in order to succeed.

Also, be prepared to be tested beyond your intellectual capacity, Snyder says, "*Being able to operate in tough times is one characteristic of entrepreneurship. Things don't always go as planned when you're developing a new company. The market isn't exactly the way you see it starting out.*"

Venture Capitalists. In a New York Times article (1), Claire Miller wrote the following. "*Venture capitalists put $7.7 billion into 1,033 deals, a decrease of 7% from the second quarter of 2008.*" She further stated that "*The number of start-ups getting financing for the first time fell 20% to 259 (in the third quarter of 2008). That is the lowest level since the first quarter of 2004. Investors are hesitant to finance new companies because they have so many portfolio companies in the pipeline that are ready to go public or be sold but are stuck because exit markets are all but closed.*" To buck this trend of VC funding, you need to have a product and an entrepreneurial team that are worthy of the investment and put in extra effort in the preparation of your business plan and executive summary.

In your search for VC funding, identify local VCs first. Check their websites to be sure that they will fund companies that are at your company's present stage (early stage, start-up stage, or first or second round of VC funding). Get a person who can link you to the VC. Participation in activities and networking of organizations like CONNECT and Tech Coast Angels not only will help make the connection, but also improve your business plan presentation. Ask your law and accounting firms for the names of persons to contact at the venture capital firms. Email solicitation may have little chance of success. If a venture capital firm invites you to make a presentation, be prepared to present your strongest statement of the market's need for your product, your competitive position, and your team's performance.

Do your due diligence on the venture capital firm and its managers. Before your interview with the firm, learn as much as you can from its website about its portfolio, companies it invests in, and its investment criteria. The firm's owners expect its managers to obtain a healthy return on the money that they have invested in it. Only if your executive summary convinces them that your company will yield a good return on an investment in your company, will they then interview you and begin the process of negotiation, due diligence and funding that is described in the following section and next Chapter. During this lengthy process, talk to the CEOs of companies in which this venture firm has invested to make sure that you will be able to work with the firm and its operating style.

Once a funding arrangement with the venture capital firm has been concluded, it will become involve in many operational aspects of your company. In addition to gaining financial support from the VC, you need to establish relations with it that will enable your company to benefit from its expertise in management, marketing and technology, instead of interfere with, the operation of the company.

The amount of capital that a venture capital firm can supply is large, leading it to become the majority owner of your company. Because the venture funding may be supplied in installments, beware of the possibility that a small initial investment might give the venture firm an opportunity to take over your company and replace your team. Don't get involved in the kind of potentially-disastrous investment deal that Snyder had concluded with a famous New York lawyer to raise funds for his mailing machine company.

Although many venture firms will count on your expertise to move the company forward, disagreements about the company's operation or how it is managed do develop. Conflicts must be resolved by negotiations that provide win-win solutions so that each party gains something and neither really loses.

Hurdle Rate. Return of Investment (ROI) is a measure of the performance of the investment. It is defined by the following:

$$\text{ROI} = (\text{Net Gain in Investment})/(\text{Initial Investment}) \quad (13.1)$$

$$= (\text{Final Invest. Value} - \text{Initial Invest. Value})/(\text{Initial Invest. Value})$$

If the ROI is provided over an investment period of n years, then the annual rate of return is

$$\text{Annual Rate of Return} = (1 + \text{ROI})^{(1/n)} - 1 \qquad (13.2)$$

If the projected Annual Rate of Return of an investment project is higher than the hurdle rate set by a venture firm, then the firm will probably consider the project.

Table 13.1 shows the guidelines that VCs or angel investors use when setting the hurdle rate for ROI. The hurdle rate depends on the company's stage of development.

Table 13.1. Hurdle Rate.

Phase I, Feasibility, Assessing or Seed Phase	80%
Phase II, R&D, Launching or Start-up Phase	60%
Phase III, Commercialization, Building-up Phase or 1st and 2nd Rounds.	40 ~ 50%

If the annual rate of return is 60%, the ROI for an investment period of five years is 949%. If an initial investment of $2 million in your start-up was made five years ago, the venture capital firm would be entitled to receive a net gain of $19 million, the value of the final investment less its initial investment. The annual rate of return is much higher than provided by the annual interest rate of house mortgages, which may be in the range of 4% to 6%. A Venture capital firm requires such an "outrageous" ROI because of the risk, which it perceives, of losing the funds that it invests in your company. Because of the confidence that entrepreneurs have in their ventures, your own risk assessment will be different from those of investors.

To better understand the relationship between hurdle rate and risk, we consider a case in which a venture capital firm makes a $2 million investment in each of three Companies: A, B and C. Company A generates the ROI given in Table 6.5, which shows an ROI of 400% or an annual rate of return of 38%. Suppose that Company B had a value of only $4.8 million, which corresponds to an annual rate of return of 19%, while Company C failed completely. The returns from these investments appear in Table 13.2. When the results of these three companies (A & B & C) are totaled in the fourth row, one sees that the initial investment of $6 million has a final investment value of $14.8 million now. This corresponds to an ROI of 147% and an annual rate of return of 20%. If company B also fails like company C did, the data shown in the fifth row (A & C & C) shows that ROI drops to 67% and the annual rate of return becomes 11 %. The actual return is even lower when management, processing and legal fees are deducted. In essence, a venture capital firm must balance losses in a large number of investment deals with big wins in a few to achieve a better ROI than available from traditional investments.

Table 13.2. Return of Investment for an investment period of 5 years.

	Value of Initial Investment	Value of Final Investment	Return of Investment (ROI)	Annual rate of return
Company A	$ 2 M	$10.0 M	400%	38%
Company B	$ 2 M	$4.8 M	140%	19%
Company C	$ 2 M	0	-100%	-100%
Companies A & B & C	$ 6 M	$14.8 M	147%	20%
Companies A & C & C	$ 6 M	$10.0 M	67%	11%

For the entrepreneurs of Company A, their stocks have risen, five years from the time the venture capital firm made the investment, to a value of $10 million. As illustrated in Tables 6.5 and 13.2, the venture capital firm's initial investment of $2 million yields a net gain of $8 million (the difference between the final and initial investment). The increase in stock value and investment value is made possible by the infusion of capital from the venture firm and the hard work of the entrepreneurs to commercialize the invention. It is the joint effort of entrepreneurs and investors that makes both winners.

Let us underscore this point by the words that Tall Oaks Capital, a venture firm in Charlottesville, VA, uses to describe itself. The firm's mission is "*partnering with entrepreneurs and management teams to build extraordinary companies.*" This venture capital firm was *formed in 2000 and is led by four partners that possess over a century of seed and early stage investment and corporate experiences.* Their website states that "*Working closely with our entrepreneurial team, we provide both the long-term investment support and the strategic guidance that seed and early stage companies need to grow and prosper.*"

13.2. Due Diligence of Venture Capital Firms

To decide whether to invest in a particular company and then execute the investment, venture capital firms undertake due diligence rigorously to ensure that the opportunity for your company to succeed is real, the business proposition is sound and the legal and financial matters are addressed. The due diligence that venture investors are likely to undertake is described below. This description was adapted from a course note that Tall Oaks Capital (4) distributed to students in a class at the University of Virginia. The process has three phases before reaching the funding stage.

Phase of Getting to Know Your Company. This is the phase during which the venture firm wants to learn more about your company when you are invited to a brief interview. The VC representatives will ask the following questions:

- Who is this company? They will want to know who the officers of the company are, as well as the names of your legal counsel for corporate matters, intellectual property, and regulatory affairs; your accounting firm and your bank.
- What does the company do? What products and services does it offer? How does your company make money? What is the size and dynamics of its market? What is the entry barrier?
- Is the company good at what it does? The VC will want to know what expertise, qualifications, and track record members of the entrepreneurial team possess. VC personnel may get in touch with references that you have provided for your team and company. They will want to know the status of your company's operations. It will be useful to present your business plan for their review.
- What does your company want from the venture firm? In addition to receiving the details of your funding request, the VC would like to know the valuation of your company, the names of any other venture capital firms contacted in this round of financing, the anticipated value of your company, the exit strategy for the investors and its timing, and your target dates for break-even and profitability.
- What are the risks in management, technology, regulatory affairs, manufacturing and marketing?

- Where will your company be in three years, five years and ten years?

Investigation Phase. Once the venture capital firm considers your company as a legitimate investment, it will commit time and resources to conduct an investigation with its internal experts, industrial consultants and university professors. Due diligence is done for the following:

- ***Product.*** This includes its functionality, benefits and advantages, expansion opportunities, proprietary position, regulatory requirements, and manufacturing.
- ***Market.*** Target population, competitors and competitive position, pricing strategy, direct or indirect sales, distribution, sales force, service and support capability, and alliances are areas to be fully investigated.
- ***Management.*** The venture capital firm will review the curriculum vitae of the principals, Directors of the Board and company's advisors. The CV should include the credentials, publications, grants obtained, positions held, and references.
- ***Entrepreneurial Team.*** During the due diligence, the VC wants to learn what communications skills, levels of enthusiasm, perseverance in entrepreneurship, optimism about the company and trustworthiness that team members possess, as well as their level of commitment. How much has the team invested in the company? Are team members proud of their ownership? Can they become discouraged by set-backs? Can they make calculated risk decision? Do they have a sense of humor?
- ***Financials.*** The VC will do due diligence on the financial condition of your company by examining its past and present budgets, past expenses for administration and R&D, future budgets, and major anticipated expenditures. They will want to know the sources of your funds and the ways you use them and, in particular, what you require to break-even and when can you achieve this. Also, what is required to generate a positive cash flow, profitability and an IPO?
- ***Exit Strategies.*** The venture capital firm will be interested in knowing the opportunity, time and valuation that will enable it to receive from your company their expected ROI for their investment.

- ***Risk Assessment.*** This will be made in areas of product and technology, corporate management, financial, market and regulatory issues.

Review Phase. This is the final due diligence. The venture capital firm will spend money to hire:

- Corporate legal counsel to review operating agreements, contracts, deeds, titles, shareholder agreements, private placement memoranda, employment agreements, etc.
- Intellectual property counsel to review license agreements, patent filings, record keeping, claims and pending claims, and intellectual property strategy.
- Technical experts to provide written evaluations, critiques and a summary of your technology, product, and service.
- CPA to assure the company's operation in accounting.

Customer references, collaboration references, and professional and personal references that you have provided will be contacted. To enable the venture capital firm to carry out its legal, financial, management and technology due diligence, you need to provide a long list of documentations in a timely fashion. In the case of a large investment, the funding may be provided by a group of venture capital firms, who will share the expense of due diligence.

Sometime during the Review Phase, the VC may send a term sheet for your signature that sets forth the key terms of the proposed investment. (See Section 14.2 for more on term sheets.) Aside from the amount established for the investment and the valuation of your company, the term sheet contains many clauses inserted by the VC for its protection, but which may reduce the value of your company. It is important to have your attorney examine and review with you each item in the term sheet. How to negotiate a deal that is good for the VC and your company is addressed in Chapter 14. Be aware that the more of your company that you give the VC now, the less you will have for later rounds of financing. Any renegotiation later may sour the relationship between the entrepreneurs and VC.

Finally, if everyone likes what has taken place during this due diligence process and reaches an investment agreement, you will look forward to signing the document and getting to the funding stage. This stage will likely begin a year after you receive the telephone call to arrange your initial interview.

After concluding the process and obtaining the funding, the entrepreneurs and the VC are now partners in running the company. At this time, you need to be sure that the closing documents and certificates are delivered to the VC soon after the signing. Schedule a first board meeting as soon as possible or at an agreed time. Updates of company progress need to be sent to investors in a timely fashion.

13.3. Account Management

Importance of Cash. Cash is money on hand or in your bank accounts. More businesses fail because of lack of sufficient cash to pay for their expenses than for a lack of profit. If you run out of cash with which to pay your suppliers and employees, you will damage your reputation and destroy their trust in you. The top priority of finance management is to have sufficient cash for bill paying at all times.

The topics covered in the next two sections are budgeting, which includes a short-term (e.g., one year) financial plan for operation of the company, financial projections, and a long-term (e.g., five years) projection of the company's revenue, cash position and/or profit and loss statement. Knowing how to manage accounts with no cash flow problems, do budgeting and prepare financial projections is important to assure smooth operation of your company.

Setting up the Account. Once the company is registered, your company can establish a bank account. You may also need to set up a petty cash account for minor transactions or purchases that require the use of cash, to aid the operation of the company.

It is a good idea to spend a few hundred dollars for an accounting program to perform the bookkeeping and financial aspects of the company. Search the internet to learn what other owners say about software programs, such as Intuit's QuickBooks. Have one of the team members do this important work and arrange to have a CPA to assist him or her in the financial work. A system that double checks the use of company funds by one or two supervisors is essential to prevent the misuse of funds, inappropriate spending or embezzlement.

The software that you obtain should be able to write salary checks for employees and pay the tax due to the government and the invoices of your vendors. At the end of your calendar year, the software will print out the W-2 form, income and expenditure statement, etc. For your company's tax filing, it is done in accordance with its fiscal year.

The financial operation should be transparent to all members of the entrepreneur team. If the team understands fully the company's financial situation, they will make the necessary adjustments to assure the company's financial health.

Make sure that you manage your accounts well so that the bank will be anxious to loan you money in the future at a favorable interest rate.

Account Management and Health of Your Company. The profit and loss statements and the balance sheets can be printed by the software for you to analyze the performance of your company as it becomes profitable. By use of the following ratios, you can determine the health of your company and better control your business's financial future (3).

Net Profit Margin = (Net Profit – Debt Interest)/Gross Sales
(The higher the net profit margin, the more profitable the business is.)

Gross Profit Margin = Gross Profit/Gross Sales
(The higher the gross profit margin, the better the business is.)

Return on Sales Ratio = Net Profit/Gross Sales
(The higher the return on sales ratio, the better the business is.)

Current Ratio = Current Assets/Current Liabilities
(Normally, the current ratio should be around 2.0 or greater.)

Debt to Net Worth Ratio = Total Debt/Net Worth
(A ratio of more than 1.0 indicates that there is too much debt for a company of that net worth.)

Cash Turnover Ratio = Gross Sales/(Current Assets–Current Debt)
(Generally, your cash turnover should be between 4 and 7.)

Average Collection Period = Accounts Receivable×365days/Gross Sales
(This should be no more than 1.5 times the permitted credit period.)

Investment Turnover Ratio = Gross Sales/Long Term Assets
(The higher the ratio, the stronger the business is.)

Return on Investment = Net Profit/Net Worth
(You should strive for at least a 12% return to have a healthy business.)

By comparing these ratios with those for previous years, you will learn whether your company is becoming financially stronger.

13.4. Budgeting

Once you have the company's finance in the coming months under control, you should start a budgeting plan for the company. You enter your company's income and expenses for the coming year into your financial planning software to determine your company's cash flow. The income might be investments made by the founders or revenues generated monthly by a consulting service, and/or commissions from the sale of other companies' products. The expenses in the budget should include salaries for personnel, purchases of equipment and supplies, rental of office and workshop, travel expenses, etc. This budgeting processing is somewhat like the operation of your household. You want to know how much you can spend when you are steadily employed and have a salary that has been committed for the coming year.

In essence, budgeting requires high certainty that you will receive the expected income, such as that provided by a contract for the sale of your medical device with payment on a certain date. If the income is based on research of the market or patient needs, you prepare a financial projection, as described in the next section, instead of a budget.

One result of the budgeting exercise is that you will know when your company will need another cash transfusion to maintain your company's "cash" situation. Another outcome could be that you will learn that your company may use up all of its cash before it can receive a transfusion. Then, adjustments need to be made to every item of expenditure. Budget cutting or revisions are always painful, but should be transparent to the entrepreneur team and some employees. Their full support of cost-cutting and their suggestions to establish a new revenue stream should be sought. Overall, the process of budgeting is a means of assuring that the company operation will experience no cash flow problems.

If your company is preparing an SBIR grant application, your software can be used to generate the budget for your SBIR grant application. The budget period for a Phase I grant is six months. It is two years for a Phase II grant. Under the current triage system of NIH, you will receive a non-funding decision from NIH in about six months. If NIH considers your proposal to be exceptional, NIH will advise you what amount of funding has been approved and when you can request the wire transfer of the funds on a certain date to your bank account for bill and salary paying. The date on which grant funds become available is usually about a year after the date of your SBIR grant submission.

If you decide to seek investment by angel investors or VCs, go to the next section to work out the aspects of the financial projection to include in the financial section of the business plan. The projections should include the income statement, cash flow statement, balance sheet, and break-even analysis.

It is possible that the process to secure angel investments will be shorter than the process of obtaining funding from the NIH. Once you receive a funding commitment from the NIH and/or angel investors, you need to revise your operation and budgeting in accordance with the funding schedule of the NIH and/or angel investors.

In summary, the most challenging aspect of budgeting is to be certain you can obtain future funding. Then, you control the expenditures by the budgeting process without experiencing a cash flow problem.

13.5. Financial Projections

Financial projections provide a big picture of the budgetary operations of your company. For a start-up or new company, it is very helpful to have projections in answering such questions as:

- What is the sale volume of your device and how much revenue does it produce?
- How many employees do you need to hire to manufacture that many devices?
- What angel investments, venture funding and/or SBIR grants are required to sustain the company's operation?
- When will be the company's operation break even?
- What will be the investors' return on investment?

The financial projections form a key part of the business plan to help investors in their decision to invest in your company. In this section, we will use a simple business model for the Excel software to prepare projections of income, expenses, cash on hand at the end of each fiscal year, value of the Company and the hurdle rate generated by the Company for a 5-year period. For the sake of simplicity, this model does not take into account such issues as tax, depreciation, bank loans, difference in timing of receipts from accounts receivable and expenditures, etc. To generate projections that include all of these issues and have fewer errors in writing your own program in Excel, you can purchase financial projection software or hire an accountant to do the calculations.

Sales and Revenues. Suppose that your Company A has three products: Diagnostic Monitor, Home Monitor and Probes to sell to hospitals for use by patients who are undergoing hemodialysis. To begin the calculations you need to make some assumption about your sale volumes. We will assume that the number of monitors sold will be relatively few during the introductory phase, increase significantly throughout the growing phase, and reach a plateau (the potential full market share) in the mature phase. The growth in sales during a period of five years is shown in Fig. 13.1, with the sales in the beginning at 3% of the full market number and the sales at the fifth year at 100%.

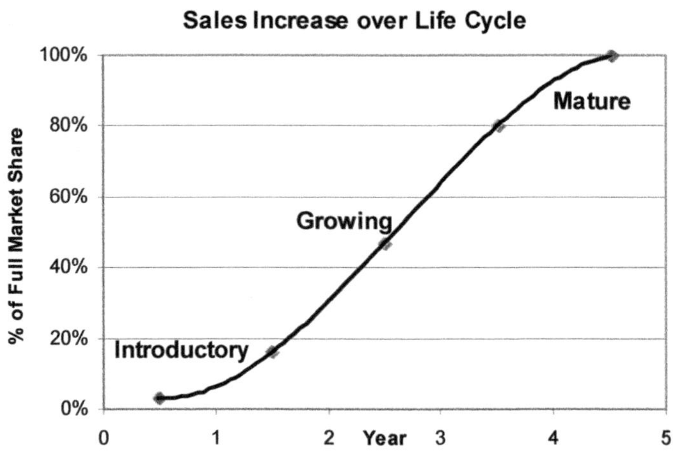

Fig. 13.1. The increase in sales during a 5-year period.

A market survey indicates that the US has some 80,000 dialyzers to serve 400,000 patients. Suppose Company A decides that the full market number of Diagnostic Monitors to be sold is 5,000 units. Then by multiplying the full market number by the percentage of the full market share (the y-axis of the growth curve) at each year, you obtain the number of Diagnostic Monitors are sold that year. This data appears in the second row of Table 13.3. Adding all numbers in that row gives the total number of monitors in use by the end of the fifth year as 12,300 Monitors. The total number of dialyzers being used is 80,000, 15.4% of which would have the Diagnostic Monitor to serve patients. (If the patient population increases at an annual rate of 5%, the use percentage 15.4% will decline to 12%.) In view of the monitor's likely benefit to patient's quality of life and a reduction in clinics' expenditures, this percentage is achievable. This suggests that the full market number of 5,000 monitors is a good assumption for financial projections.

Similarly, Company A sets the full market number of Home Monitors sold at 20,000 units. The numbers of Home Monitors sold each year are shown in Table 13.3. The total number of monitors sold during the five years indicates that 12.3% of all hemodialysis patients would be using a Home Monitor for their home dialysis. This percentage is achievable in five years in view of the two following observations (2). First, 0.8% of dialysis patients in the US were on home dialysis in 2003, but about 30% are suitable candidates for home dialysis. Second, almost 55% of dialysis patients in New Zealand were already on home dialysis in 2008 (2).

The full market sale numbers are used as the inputs to the Excel program. They can be altered if further market studies are undertaken.

Because of the wear and tear on the probe, customers who use the monitors need to replace the probe each year to ensure its functionality with the monitor. The Excel sets the total number of all monitors in use at the end of the first year as the number of probes being sold during that year. The total number of probes sold each year is given in the last row of Table 13.3.

Table 13.3. Projected Numbers of Monitors and Probes sold by Company A.

	Year 1	Year 2	Year 3	Year 4	Year 5
% of full market share	3%	16%	47%	80%	100%
Diagnostic Monitors Sold	150	800	2,350	4,000	5,000
Diagnostic Monitors Accumulated by year end	150	950	3,300	7,300	12,300
Home Monitors Sold	600	3,200	9,400	16,000	20,000
Home Monitors Accumulated	600	3,800	13,2000	29,200	59,200
Probes Sold	750	4,750	16,500	36,500	71,500

The next task is to use the market survey of Company A to set the prices for the two monitors and the probe. Suppose that competitors are selling monitors at a price of $4,000 each. Although the Diagnostic Monitor may perform better, the Company sets its selling price at $4,000. Because the Home Monitor is simpler, its unit selling price is set at $2,500. In view of the expense to manufacture the probe, its unit selling price is $400. With the numbers sold and this pricing scheme, the Excel calculates the revenue that appears in the top part of Table 13.4. For the full market numbers used, the revenues of Company A are projected to reach $95 million by end of the fifth year.

The contribution to revenues by product categories is depicted in Fig. 13.2. The lowest curve illustrates the revenue from the sale of Diagnostic Monitors, the middle curve traces the revenue from the sale of Home Monitors, and the highest curve represents revenue attributable to sales of Probes. Although only 12% of dialysis patients are using Home Monitors, their sales make up 53% of total revenues of the Company at Year 5. The figure also indicates a faster growth rate in revenues from sales of probes as disposable items.

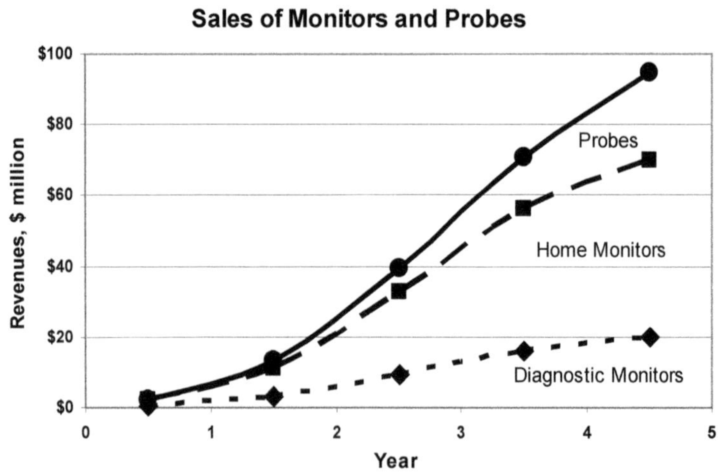

Fig. 13.2. Revenues of Company A generated by the sales listed in Table 13.2.

Expenses. As shown in the second segment of Table 13.4, seven categories of expenses are considered when determining the total expenditures of Company A. From the knowledge gained in manufacturing prototypes, the material costs for making Diagnostic Monitors, Home Monitors, and Probes are estimated at $500, $400 and $150 respectively. The Excel program combines the costs of calibrating and testing the probe, labeling the monitors and packaging, insuring and shipping them and servicing as "manufacturing, packaging and shipping expense." It is estimated that this expense represents 55% of material

Table 13.4. Income from sales of three products, Expenses for Administration, R&D and others, EBITDA, Ending Cash, Net Company Value, Return of Investment, Hurdle Rate of Company A. (All are in $ millions)

	Year 1	Year 2	Year 3	Year 4	Year 5
Income					
Sales of Diagnostic Monitors	$0.6	$3.2	$9.4	$16.0	$20.0
Sales of Home Monitors	$1.5	$8.0	$23.5	$40.0	$50.0
Sales of Probes	$0.3	$1.9	$6.6	$14.6	$24.6
Total Revenue	$2.4	$13.1	$39.5	$70.6	$94.6
Expenses					
Administrative Overhead	$0.4	$0.9	$2.2	$3.8	$5.0
Overhead for R&D	$0.4	$0.9	$2.2	$3.8	$5.0
Infrastructure & Rentals	$1.5	$1.3	$4.0	$7.1	$9.5
Materials	$0.4	$2.3	$6.6	$11.6	$15.6
Manufacturing, Packaging, Shipping & Servicing Expenses	$0.2	$1.2	$3.6	$6.4	$8.6
Marketing & Selling Expenses	$1.1	$5.9	$17.8	$31.8	$42.6
Bad Accounts	$0.1	$0.5	$1.4	$2.5	$3.3
Total Expenditures	$4.1	$13.0	$37.8	$66.9	$89.5
EBITDA	-$1.7	$0.1	$1.7	$3.7	$5.1
Beginning Cash: $2.0 M					
Ending Cash	$0.3	$0.4	$2.1	$5.8	$10.9
Net Company Value			$8.6	$18.4	$25.4
Amount Paid to Investors			$4.3	$9.2	$12.7
ROI			115%	360%	535%
Hurdle Rate			29%	46%	45%

costs. We assume there are two overhead expenses. The overhead incurred by the management team and the board of directors has a fixed expense of $250,000 plus 5% of the total revenue. The same amount is set aside for R&D. A similar arrangement has been made for the expenses incurred for infrastructure and rentals. Because of the need for a large investment in equipment and facilities, we will use a much higher infrastructure expense for Year 1. The marketing and selling expense to conduct the marketing campaign, digital marketing, and workshops has been estimated at 45% of total revenue. Finally, uncollectable accounts receivable are projected at 4% of revenue. These are sales on credit that are not collectable. Total expenditures are shown on the last row of the second part of the table.

EBITDA and Ending Cash. The difference between revenues and expenditures constitutes the net earnings. For the current income and expense computations, the earnings are known as EBITDA (earnings before taxes, depreciation and amortization. Because of the large expense of infrastructure, administration and R&D, we see negative earning in the first year. With the initial influx of $2 million from the investors, the cash that Company A has at the end of first year is about $0.3 million. Because of the positive earnings projected for Year 2, the ending cash for that year is slightly larger than for Year 1. The values of EBITDA and ending cash during the five years are listed in the third segment of Table 13.4.

In this table, the smallest ending cash is $0.3 million. In the yearly financial projections, this ending cash serves as the cushion or working capital to prevent a negative cash flow caused by the time lag between expenditures and payments of account receivables. Commercial financial software can refine the projections to show whether there will be a cash flow problem if data are calculated on a monthly basis.

Company Value, ROI and Hurdle Rate. Suppose that the investors agree to use 5 as the EBITDA multiple to calculate the net company value, which is in the first row of the fourth segment of Table 13.4. Since the $2 million investment gets the investor 50% of the company, the ROI calculated to the end of Years 3, 4 and 5 are provided in the third row of this last segment of Table 13.4. The hurdle rate is given in the last row. The hurdle rate for Year 4 and 5 are comparable to those established by VCs for companies in the building-up phase (Table 13.1).

Potential Saving for Clinic. The total amount paid by clinics to the company in Year 5 is $94.6 million. There are 110,000 patients using the monitors. Thus, the expense incurred by clinics to acquire the monitors and probes for patients use is $855 per patient.

Suppose the annual charge for hemodialysis is $60,000 per patient. Because more patients now use home care and have a lower frequency to develop intradialytic hypotension, it is appropriate to suggest that these improvements produce a 7% reduction in clinics' operation expense (i.e., $4,200). Because healthcare givers need to spend time to administrate the monitor, the labor cost is estimated as 3% of the annual charge for hemodialysis (i.e., $1,800). When the difference between the reduction in operation expense and the increase in labor cost is taken as the overall operational saving (i.e. $2,400 = $4,200 - $1,800), one sees that the use of monitors to improve the delivery of hemodialysis might mean that clinics that spend $855 per patient to acquire the monitor will gain an operation saving of $2,400 per patient. This is a ROI of 181% based on data in Year 5. (A verification of these saving projections by actual records should be sought in years earlier than Year 5.)

Limitation of Financial Projections and Sensitivity Analysis.

Once you do the projections, ask critics from the entrepreneur team and outside financial experts to revise them. In the current business model, the overhead is a base expense plus a percentage of total revenue. Check these overheads against the organization structure of the Company to see whether the overhead is sufficient to pay the salary and fringe benefits of executives and their administrative assistants who appear in the organization structure. Similarly, check the estimates of the manufacturing, packaging, shipping and servicing expenses against the number of employees required to do the manufacturing, assembling and packaging jobs.

Don't overstate your market share. Overestimation, underperformance, and financial crisis can always occur. Having sufficient ending cash or working capital and strictly controlling your expenditures are essential for your company's safe sailing.

Finally, have fun playing with the numbers (i.e., do a sensitivity analysis on the values used as input) so that you know what impact those numbers will have on the financial future of Company A.

References

1. Miller, C., bits.blogs.nytimes.com/2009/01/05/
2. Rubin, R., USA Today, http://www.usatoday.com/news/health/2009-08-23-dialysis_N.htm?POE=click-refer (2009).
3. Sitarz, D., *Simplified Small Business Accounting*, Nova Publishing, 1995.
4. www.talloakscapital.com

Chapter 14

Negotiating Deals

14.1. The Negotiation Process

The Smart Negotiation. As a CEO, you will negotiate with a VC firm for funding, an executive to employ, and another company about joint licensing, a merger or an acquisition. The beginning of a round of negotiations or bargaining is competitive as each side will have different values or goals on funding, employment, licensing right, fees, or price etc. However, in the end, the negotiation of a good deal that can lead to years of collaboration and terms that satisfy both sides can be termed smart negotiation. A good CEO must have the skill to carry out smart negotiation for important actions, such as VC funding, executive employment, and joint venture development.

Be sure to do your homework and prepare for your negotiations. Think about your opponent and yourself. Work on the answers to these questions:

- What do you want?
- Does each party meet the other's requirement?
- What do you expect to achieve from the negotiation?
- What is the bottom line position of your expectation? What is the least that you will accept and the most that you will concede?

During negotiation, you will learn more about the other side and its position and gain an idea of the likelihood on being able to come to a realistic outcome. Use the following pointers to help you complete your smart negotiation:

- Negotiate in a "win-win" mode for competing, but collaborative, deals. Don't attempt to out-negotiate the other side.
- Work toward a deal. If there is no deal, what are your alternatives?

- Don't get hung up on insignificant issues. Look for trade-offs or alternatives.
- Be creative and flexible, especially if the deal gets "hung-up".
- Bargain with give and take, so that both sides feel that they are getting something extra.
- Be prepared to give concessions. Ask for more, or swap, on issues. You should not win them all. Leave something on the table for the other side.
- Never let yourself be verbally abused by the other side. Use questions to deflect abuse.
- Strive for credibility in your reasoning and compromise.
- Let your strength speak for itself.
- Ask for clarification as a means to justify your position.
- Never use threats. Gently and with great care to set a deadline.
- Uncover information that may benefit your negotiation.
- Understand the power of questions and the power of silence in negotiations.
- Take control of the negotiation by having a realistic proposition.
- Maintain good negotiating judgment with balance and moderation, and avoid extreme positions.
- Know when and how to wrap up a complete package deal.

The Hostile Negotiation. A few months after your medical device has appeared on the market, you may receive an injunction from your competitor that demands that you to stop selling your device because it infringes upon many of their patents. If you fail to do so, your competitor will probably file a lawsuit in the District Court to sue you for damages and a punitive penalty. The receipt of the injunction letter marks the beginning of hostile negotiations and a costly lawsuit. Aside from your competitors, it is also possible that a patent troll may sue you for infringement of its patents. A patent troll is a person, company or a non-practicing entity that owns patents, but has no intention of manufacturing or using the invention, but wants to sue aggressively in the hope that it will receive licensing fees (2).

Here are some of the reasons why you may be sued if you market your invention:

- **To eliminate you as a competitor.** Your competitors don't want you to compete with them as your success may put their non-competitive products out of business. The legal fees for a medium-size infringement suit can be $2 to $5 million. A small start-up may be sued until it can no longer pay an attorney to defend it. Without a good legal defense, you can lose the suit and put your company in jeopardy.

- **To obtain a licensing fee or payment for a settlement**. In 2001, the Israel Company, Medinol, initiated a lawsuit against Boston Scientific, accusing the latter of stealing Medinol's stent-manufacturing technology. Boston Scientific settled the lawsuit in 2005 by agreeing to pay $750 million to Medinol (1).
- **To merge with, or acquire, your company.** In some instances, this implies that a hostile negotiation will become a friendly negotiation so that both companies can benefit from the merger or acquisition.

Such a legal challenge will immediately cause trouble as you are attempting to raise money for the company. Although your product may be promising, your potential investors may withdraw to the sidelines because of the uncertainty of outcome of the lawsuit and the expense that will be incurred to defend it. Your company will be in even greater distress if you do not put in the effort and money to defend your company. If you do, you may end up doing one of the following:

- **Settle before anyone files a legal suit.** Suppose that the due diligence of your adversary's patent reveals that your product does not infringe upon the patent. However, if there is a possibility of infringement, you discover that you can design around your opponent's patent so that your new product will no longer infringe upon the patent. After your patent attorney presents these arguments to your adversary, it may feel that the outcome of its legal action is unlikely to yield a net payment and thus decide to settle for nothing or for a small payment.
- **Settle before trial.** Once a lawsuit has been filed and deemed to have merit by the judge, he or she will set a trial date. However, about 90% of the cases are settled before the date of the trial. Before the trial date, there is a discovery stage in which both plaintiffs and defendants must disclose everything they know that is related to the legal case. The information discovered at this stage may indicate that the opponent's patent is invalid and that there is no infringement or that there is substantial

infringement. Even in the latter case, it is not certain that a jury verdict will favor the infringers or infringees. Consequently, you and your opponent may decide to settle the case before the trial. Sometimes, the judge and the two parties may request that a mediator be appointed to evaluate the case and guide the two sides in working out a settlement to which they mutually agree before the trial.

- **Settle by trial.** The attorneys will present their cases to a jury for a verdict. The overall legal costs will exceed the two previously mentioned settlements. The verdict will include damages (mostly loss of royalties), compensation for punitive damages, and the fees of attorneys for the winning side. For example, the Tel Aviv-based Medinol Company sued Boston Scientific in 2001 on a drug diluting stent technology in a Dutch court, winning a $750 million settlement in 2005. But that case wasn't closed until 2011, when the companies agreed to cancel some provisions of the 2005 settlement and release each other from all claims in the case. Medinol paid Boston Scientific $104 million as part of that deal in January 2011.

Historically, U.S. patent owners have been entitled to an injunction upon a finding of patent infringement. Since the ruling in the case of *e-Bay v. MercExchange* by the Supreme Court, several District Courts have refused to grant injunctions to patent trolls (1). These court decisions will have a profound impact on patent licensing as these precedents remove the use of an injunction by a patent troll to stop an infringer from making and selling infringing products. As a result, the strength of a patent troll in licensing negotiations is diminished, the bargaining power of the accused infringer is enhanced, and the "infringing" company can better afford to engage in litigation. In addition, District Courts have granted compulsory licenses to companies found guilty of patent infringement in some cases, making a patent troll more willing to grant a license to an accused infringer.

Dispute Resolution. Arbitration is stipulated as the means to resolve disputes in many employment and user agreements. For this technique, the dispute is referred to a third party to review and to impose a settlement decision that is legally binding. Arbitration is most commonly used for the resolution of commercial disputes. Often the fine print in agreements contain a stipulation that arbitration be used to settle disputes arising from the agreement and deny the use of the courts by employees and consumers as a means to settle disputes. If a dispute is highly technical, an arbitrator who has the appropriate expertise may be appointed. In general, arbitration is faster and cheaper than litigation in court.

Mediation is another technique used to resolve disputes. Most judges will recommend that the litigation parties go through the process of non-binding mediation before proceeding to a jury trial. Many mediators are former judges who know the outcomes of cases that were similar to your case. Mediators use their knowledge and skill to improve the communication between disputing parties and to help them reach an agreement. They are very experienced in finding a middle ground to which both parties will be able to agree. The mediators will advise all parties that this middle ground will definitely upset them, but is the least objectionable option in view of the high legal expense of proceeding to a trial and risking its uncertain outcome, with or without any monetary award.

14.2. The Venture Funding Term Sheet

Overview. If a VC firm invites you to an interview, you will present your business plan to describe what you are offering and what funding you are looking for. Sometimes during the negotiations, the VC will let you know that they are interested in financing your company and will send you a term sheet that sets forth their key terms and conditions of their investment in your company. This is a non-binding agreement and some terms and conditions are open to negotiation. If you sign the term sheet, you are indicating that you will limit your funding negotiations to this particular VC. The term sheet may stipulate that you pay the legal fees of the VC firm's legal counsel that are incurred by that firm in its due diligence of your company if you reach agreement on the funding arrangement.

Before getting to the subject of negotiation, let us examine how the VC operates its business. A VC firm raises fund from institutional investors, such as pension funds of state governments or large companies on the strength of its managers' reputation. Suppose that the investors invest \$100 million in this VC firm and set a term of 10 years as the period in which it will conduct its business. Even though the management fee that a VC firm usually charges is 2 to 2.5% a year, an overall management fee of 2% for 10 years will provide a \$20 million fee to the VC firm. If this is set aside, the amount available for investment is reduced to $80 million. If the $80 million are invested evenly over a five-year period, the VC will have $16 million for venture investments for each of the five years. VC firms normally spend the remaining years of the fund's life in managing those investments. A VC firm raises a new fund every three to five years, so that it can continue to actively invest in ventures. The long term health of a VC firm can be determined by how recently it has raised new funds.

Because of the economy in recent years, it is harder to liquidate or maintain the viability of the companies in a VC firm's portfolio. As a result, lower VC funding for new companies creates a buyer's market for VC firms that enables them to be selective when granting an investment deal. Thus, a VC firm will tend to set a lower valuation on the company, so that the same investment can obtain for it a larger ownership fraction of that company. If you are able to work out funding to sustain the operation of your company, you are in a stronger footing to negotiate a better funding deal.

Company Valuation and the Price of Preferred Stock. One of the most challenging aspects of negotiation with a VC firm is to reach an agreement on a valuation of your company. You may use the market potential of your company or its revenue and earnings to justify a valuation. On the other hand, the VC may use the track record or success of your entrepreneur team to arrive at another valuation. Like the process of selling a house, only after the VC and you have agreed to a value, will you proceed to negotiate other conditions in the term sheet.

Angel investors and the company's founders usually receive common shares in exchange for their investments. A VC wants to receive preferred shares, which have priority over common shares in liquidation and the distribution of assets, as well as voting rights that provide greater protection for the VC and full control of the company. The VC's choice of preferred stocks is not negotiable. However, some of the protection sought and the company valuation, which is used to determine the price of preferred stock, may be negotiable.

Suppose the VC will invest $2 million in Company A as an agreement on the valuation of Company as $2.6 million is reached. In view of the numbers listed in Table 6.4, this valuation is equivalent to set a price of $2.00 for each share. Thus the VC would receive 1 million shares for their investment. However, because the VC will receive preferred shares, it can stipulate in the term sheet that it will have control of the Board of Directors, even though it is not the majority owner of Company A as it does not own a majority of the voting (common) shares. After the VC investment, the valuation of the company will become $4.6 million.

Preference in Liquidation. A liquidation may be caused by an acquisition of Company A by others, a sale of the company, or a bankruptcy proceeding of the company. The VC has preference in the distribution of cash generated by such a liquidation. This preference can take the form of a return to the VC of an amount that is equal to the amount that it had invested, twice the amount that it had invested or three times its investment, depending on the risk of investing in Company A. If the preference is 2X and Company A is sold for $4 million, the VC will receive all of the $4 million, leaving nothing for the other shareholders.

If the selling price of Company A is $7 million, then the additional $3 million will be distributed to the other shareholders, according to their ownership shares. This double dipping, the preference distribution and the subsequent distribution to the other investors, should be negotiated.

Dividends. This is the annual return to shareholders for their investment. For a company in the first or second round of VC funding, the dividends are usually shares, instead of cash. The dividend to preferred shareholders may be equal to a predetermined and fixed percentage of their investment. Sometimes the term sheet permits the Board of Directors to vote on the dividend's percentage of return. This may be lower for common shares. Try to negotiate that any dividends to be paid will be non-cumulative. If the annual return is set at 8%, the total non-cumulative return for an initial investment of $2 million over five years is $800,000. If it was compounded, the cumulative return would come to $939,000.

Redemption Rights. One condition stipulated by the VC might be that Company A will redeem the preferred shares in three annual installments, beginning on the closing day of the fifth year and at a price equal to the original purchase price (i.e., $2 million) plus the accumulated dividends (i.e., $800,000). This condition is designed by the VC as an exit strategy when the company is progressing well, but is not yet a candidate for an IPO or acquisition. This redemption right is designed by the VC to correspond to the life span of venture funds.

Board Composition. The term sheet may establish that the holders of preferred shares can choose four Directors to sit on the Board, that the holders of common shares can choose two Directors and that you and the VC will agree on the choice of one independent, additional Director to fill the Board position. Although the holders of preferred shares have fewer shares than do the holders of common shares, the preference enables the former to effect control of the Board and subsequently the company. This control could result in the appointment of a new CEO immediately following the investment is concluded.

Drag-along Rights. If the holders of preferred shares decide to liquidate and sell the company, the clause on drag-along rights requires the holders of common shares to agree to the liquidation or sale. Suppose that the preference is 2X (i.e., $4 million) and the dividends due to preferred shareholders are $800,000. If the sale of Company A generates cash of $4.8 million, the founders and the holders of common shares will get nothing.

Reverse Vesting and Key Employee Matters. Before the closing of the investment deal with the VC, the founders have one million common shares. After the closing, the founders must set aside their common shares and then earn them back over the next four years. This stock redistribution process is termed reverse vesting. You may negotiate to immediately obtain a portion of the shares (this is also known as cliff vesting) and/or shorten the vesting period to two or three years. One of the reasons to have this condition in the term sheet is to ensure that the founders stay around and help the company to develop and progress. In addition, a certain portion of stock is specified in the term sheet as a set-aside for an employee stock option plan.

Anti-dilution Provision. This protects investors when shares of stock are sold at price per share less than the price paid by the investors. Preferred shares are normally convertible at the option of their holders to common stock on a share-for-share basis. Price-based anti-dilution adjustments involve increasing the number of shares owned by investors.

Registration Right. This is a right given to investors to demand that the company register its shares with the SEC to enable the investors to be able to sell their shares.

Pro-rata Participation Right. This right entitles investors to participate in future equity offerings on a pro-rata basis.

For more information on the topics introduced or discussed here, consult a competent attorney or CPA who specializes in venture investment.

14.3. Executive Employment Contract

Overview. This contract records the terms of employment and compensation for the executive, the rights of both the executive and employer, and the company's obligations to the executive. As the CEO, you need to work out an employment agreement that will motivate the new executive to become a team member without costing the company a fortune. Even if you are the sole founder of the company, you still should have an employment contract for yourself to ensure that your hard work and contribution will be rewarded and that your legal rights are protected.

Your recruitment pamphlet will describe what qualification and knowledge are sought in the executive being recruited. You can ask a placement firm to help you in the search for the right candidate. Many entrepreneurs volunteer their time to help the advisory work offered by SCORE or organizations, such as CONNECT. These are sources for you to use to find executives to serve in your company.

It is important to do due diligence of the new executive before the interview and signing the employment contract. Knowing the professionalism and characteristics of the executive and presenting a legally enforceable employment contract right in the beginning will save many disputes, headaches, and remedial actions. In one case, a company was happy to recruit an executive to work on some patentable inventions. He was granted a significant number of company shares. Soon afterward, the partners discovered that this executive also worked part time for a competing company on the development of a device of a similar nature. Although this ethical violation of the executive's employment terms enabled the partners to eventually fire him, hiring him had already put the patentable work done by the company in jeopardy. Because of the lack of a carefully written employment contract, additional effort had to be spent to recover the fired executive's company shares.

The contract for the Chief Executive Officer (CEO) should include the following provisions:

General Description of the Position, Responsibilities and Term. This spells out the CEO's position. You need to set out the duties and responsibilities clearly in the CEO employment contract. It may state whether you will have a seat on the Board of Directors and how you will report to the Board of Directors and the Chairman of the Board. It may limit the activities that the CEO can engage in, such as consulting work, business activities outside of the company, employment by another company and ownership of an interest in a company that is in competition with your company. The CEO's term of employment is usually set at five to ten years, but is negotiable.

Compensation. This includes salary, annual bonus, stock options, employment benefits, expenses, vacation and other perks and/or a golden parachute.

Trigger Events. The contract will define what will happen to you if there is a change in control of the company, as occasioned by a change of ownership, sale of the company, or merger or acquisition of the company. Your position as CEO is an at-will position, meaning your employment can be terminated at anytime for cause, without cause, due to disability or to death, for good reason or without good reason.

Proprietary Information Agreement and Termination Obligations. The agreement deals with the retention and ownership of any intellectual property by the company, the non-compete nature of your future job, the protection of the company's trade secrets, the prohibition of the solicitation of the company's customers and employees, the return of property and documentation, and assistance in litigations when you leave the company.

Arbitration to resolve conflicts, the release of claims, the payment of legal fees and costs and the law governing employment contract will be defined in the contract.

The following issues may require more scrutiny by the executive and his/her attorney:

Stock Options. Stock options are employed for the company to attract and retain talented employees, help motivate more dedication from employees, to be a cost-effective employee benefit plan, in lieu of additional cash compensation and to help smaller companies compete with larger companies in attracting great employees. If it is a restricted stock grant as used in many start-up companies, the executive will be required to pay taxes on the fair market value of the stock on vesting. If the stock option is properly structured under an "incentive stock option" plan, an executive will not have a tax payment to make when his shares are issued, but will pay regular taxes when the stocks are sold.

Definition of Trigger Events and the Compensation. These events not only can lead to the termination of employment contract, but also determine the amount of compensation. A clear definition of the events is necessary for your protection. The form of compensation may include accelerated vesting of stock options and/or continued payments of salary, bonus and paid medical benefits. The amount and timing of these payments may have tax consequences. These issues should be addressed in the employment contract at the outset to avoid additional tax penalties.

Intellectual Property Ownership and Other Restrictive Clauses. As an executive, you may have an interest in the development of ventures outside of the area in which the company competes (or will compete). The right to continue to involve yourself in such ventures and to retain ownership of your inventions should be negotiated. In addition, the restriction on post employment solicitation of customers and employees and non-competition may need to be examined in order to avoid costly litigation after termination of employment.

14.4. Licensing Agreement

Overview. One objective in granting a licensing right for an intellectual property is to facilitate the joint development of a product. Licensing can be the first step in forming an alliance or the establishment of a joint venture, or the prelude to a merger or acquisition. You may use the licensing right as a means of entering a new area of technology. Most licensing agreements are created collaboratively, while others are coerced by litigation or the threat of IP infringement.

The company that grants the licensing right is known as the licensor and the one that receives the right as the licensee. A licensing agreement is written with these objectives in mind:

- To define the economic and technical relationship between the licensor and licensee
- To assure the rights and procedures on the use of the licensed IP
- To describe the ways by which the use of the licensed IP will be monitored, regulated and terminated.

Issues in the licensing agreement to be negotiated include:

Licensing Rights. Once you receive a patent from the USPTO, you can exclude any other persons or companies from making, using, selling, or offering to sell in the USA, or importing into this country in which the patent was granted. The licensing right allows the licensee to carry out these activities without exclusion by the patent owner (i.e., the licensor). If the invention is associated with more than one patent, continuation-in-part or patent application, the licensed patents are referred to as the bundle of these patents. For trade secrets, a licensing right allows the licensee to use the secrets and to access information from the licensor. For copyrights, the right granted by the licensor enables the licensee to copy, reproduce, distribute, publish, create derivative works, display and perform in public that which was copyrighted. Licensed technology, referred to as unpatented technology, includes documents and biological materials, software codes, etc. Its licensing is subject to the rules established for trade secrets and copyrights. The licensee should do due diligence on the validity of licensed patents and technology and ask the licensor for full disclosure. Trademarks for certain products can be licensed for use by means of a licensing agreement.

Licensed Rights and Products. Most products will undergo the following development process: R&D, manufacturing, regulatory evaluations, marketing, distribution and servicing. The licensor can grant this right for all of these processes or to only some of them, as outlined in Fig. 14.1. As an example, the licensor can decide to license only a right to manufacture the device to the licensee. In this licensing or outsourcing arrangement, the licensee will deliver the product to the licensor and the licensee will not have the right to carry out other activities, such as selling or distributing the device.

A licensed product is one that is made by using the licensed patent or technology. It should use a label that identifies the patent numbers or shows the term, "patent pending." A careful definition of the number of units of the licensed product (e.g., medical device) that are produced or of the way in which the software will be used forms the basis for calculating the compensation to pay the licensor.

Territory. The territory for which the license will be applicable can be negotiated. It may be the USA and its territories or simply one or more specified states.

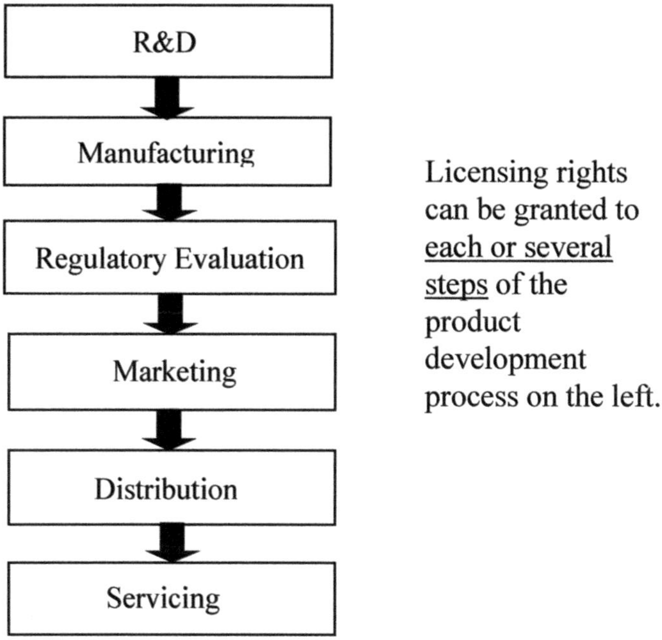

Fig. 14.1. The product development process and its connection to licensing rights.

Field of Use. This defines the type of products and processes that can use the licensed technology or product. A disease may be specified when defining the field of use. For example, if a patented device is licensed for use in neural diseases, the licensee cannot use the device for use in cardiovascular diseases. Conditions can be spelled out for the use of licensed technology in research and non-commercial uses.

Compensations. If a right is granted to the licensee to sell the licensed products, the compensations are negotiated. This may include the payment of upfront fees for the transfer of the technology and installment payments according to certain developments or attainment of marketing goals. Royalty payments are appropriate. They can be calculated on the basis of the number of units or amount produced or sold. Payments can be based on gross sales, net sales, or net profits. The terms for the calculation of compensation should be clearly defined and the licensor should be given the right to audit and verify the calculation of payments and the figures on which they were based. The manner of payment (by cash, payment in-kind, or product transfer) and its timing need to be spelled out in the licensing agreement.

Sublicensing. In contrast to general contracts, licenses are not assignable by the licensee without the licensor's consent, unless the license agreement explicitly states that the license is assignable. When control of the licensee changes hands, the license is not assigned to the licensee's new owner. On the other hand, the sale of the licensee's shares will not affect the license agreement. Sometimes, an ability to transfer the license is desirable, if its transfer to a new owner will enhance the valuation of the licensee. Then, transferability should be negotiated.

In addition to the issues covered here, the way to transfer and escrow the licensed technology and/or materials used for it, the law governing a license agreement, prosecution for license violation and dispute resolution should be specified in detail in the licensing agreement.

References

1. Heuser, S., *The Boston Globe*, Sept. 22 (2005), www.boston.com/business/technology/biotechnology/articles/2005/09/22/boston_scientific_medinol_settle_row/
2. Lerer, L., www.law.com, July 20 (2006).
3. Mauro T., www.law.com, May 16 (2006).

Chapter 15

Leading People and Managing Yourself

15.1. Preparing Yourself and Managing Your Time and Work

The CEO and his/her entrepreneurial team are the heart of a company. Raising funds for the operation of the company is a full-time job. Leading people and managing the team is another full time job. It is important that you divide your limited time between these two jobs, while maintaining a balance between your business and family life. Since your idea of the entrepreneurial venture will have a limited useful life, so will you. To succeed, you will have to constantly change and adapt. Although you may change your business ventures, your family will be with you for many years to come, especially after you have retired from your entrepreneurship.

You need to constantly look for opportunities to create for your company. Finding potential acquirers with financial strength will get your company over and pass the Building-up Phase. Examine your company's position to find ways of generating revenue. The best position you can put your company in is having sufficient money in the bank to pay your vendors and employees without the need for VC funding.

In Chapter 3, we identified a number of traits that an entrepreneur should possess or acquire (Fig. 3.1). Here are a number of rules or bits of practical wisdoms for you to consider for effective, day-to-day operation of the company:

***On Interpersonal Relationships*:**

- Be truthful in your dealings with your team, your customers, your family, and even your competitors. Above all, however, be truthful to yourself.
- Be honorable and ethical in your business dealings. You will at times be confronted with a deceitful person or company. Learn to recognize deceit and how to deal with it. Avoid direct confrontations.

- Negotiate in a "win-win" mode. Remember that you will often be creating a long-term relationship with "the other side."
- Establish business friends, associates and future contacts by applying the three rules above. In hard times, these contacts will stand by you. Don't get dropped like a hot potato when the going gets tough.

On Productivity:

- Plan your day and set priorities. Maintain a "to-do" list. Work on high value tasks and get them done on time. Stay on top of your most important responsibilities; the others will just have to wait.
- Define goals and problems clearly. Clarity is everything. When discussing a problem, focus on the future instead of the past. Deal with problems creatively and effectively.
- Always be open to feed-back and listen to negative viewpoints.
- Respond to your co-workers' messages and requests. Clear and timely communication saves time and reduces any anxieties that your co-workers may have. Create notes before making calls and keep them brief.
- Use a system to keep track of names and phone numbers.
- Break up your day into units, for example units of 75 minutes with a 15-minute break between. Use the break to return emails and phone calls, hold quick conferences, and get coffee or tea.
- Take one day a week off "just for you and your family".

On Mastering the Internet:

- Learn how best to find useful information online.
- Organize and manage information flows that are somewhat overwhelming.
- Network on the internet to accomplish professional goals.
- Leverage internet tools to accomplish results more quickly.

On the Company's Operation:

- Concentrate on making sales and controlling costs. Learn both and do them well to achieve maximum profit for your business.
- Evaluate expenditure before you make it without exception. Always look for a less costly alternative. Maintain the habit of conserving cash and reducing the costs of everything you do.
- Hire employees after:
    - Reviewing whether you can afford to hire an employee.
    - Identifying the type of employee needed.

- o Defining the work to be done by the employee.
- o Recruiting and interviewing prospects.
- o Familiarizing yourself with the legal requirements governing the employment relationship.
- o Complete all paperwork that is necessary when hiring an employee.

On Self Improvement:

- Read everything and attend every seminar and workshop you can. You will learn new ideas, techniques and methods to save you hours and days of hard work and research. Seminars and workshops are good places for networking.
- Think out of the box, aggressively seek new ideas, question your assumptions, and ask for accurate information. There are dangers in conformity and complacency.
- Learn the know-how of running a company. 80% of businesses started by experienced business persons succeed. The reason is that they have learned how to purchase, negotiate, raise money, sell and market, manage the finances and lead employees.

15.2. Setting Goals, Incentives and Recognition for the Team

As the leader of your company, you will need to communicate well with your team, develop consensus among team members, delegate jobs and authority to the team, set SMART goals and establish incentives. Make the decision process transparent so that your team will not become discouraged with rumors from other sources. As you delegate work to your team, stay on top of the situation and have frequent meetings with the team to monitor progress. Give employees the authority to do their jobs, but not treat your delegation as a means to abdicate your responsibilities.

The SMART goals to be set for the team should have these characteristics (4):

- ***Specific.*** The goals must be in writing and very specific. A specific goal is much easier to understand and has a greater chance of being accomplished than a general and vaguely-worded goal. You must specify:
    - o Who will do it.
    - o What you want to have accomplished.

- Where you want it done.
- When it should be completed
- What the support, requirements and constraints for completing the goal are.
- Why you want to achieve the goal and what the purpose and benefit of accomplishing the goal are.

Avoid non-specific statements, such as "to always work for the good of the company." A team member should only know what is required for his specific assignments.

- ***Measurable.*** Establish a time to reach the goal and specific criteria for measuring progress toward attainment of the goal. To determine whether your marketing goal is measurable and quantifiable, ask questions, such as, "How many devices do I want to sell in the first quarter?" or "By when do I want this goal to be achieved?"
- ***Attainable.*** Specify the action to attain. Your team's attitude, abilities, skills, resources and financial capacity make it possible to reach the goal. If you plan your steps wisely and establish a time frame in which to carry out the steps, you can attain almost any goal that you set. When your goals are attainable, you improve not only your self-image, but also your self confidence.
- ***Realistic.*** There must be a possibility that the goal can be reached. To be realistic, a goal must be something that your team is both willing and able to do. A goal can be both high and realistic. A high goal is frequently easier to reach than a low one, because the former will require greater effort - and motivation - than the latter. Do the job as a labor of love.
- ***Timely.*** Timelines must be set within a context. If there is no deadline attached to the goal, there will be no sense of urgency.

To begin developing effective team effort right, adopt the habit of setting business goals daily, weekly, monthly and quarterly, putting them down in writing daily and weekly, and monitoring them daily. Because of Murphy's Law, allow at least 50% more time than you feel is needed for a task. Innovation, R&D and ventures always are more expensive and take longer than expected.

Make sure that every team member knows that time is truly money. Similar to goal setting, a company must have ***a SMART incentive plan*** to recognize the goal accomplishers of the team. When your company's profit exceeds a certain specific level, then award a measurable number of dollars as an incentive or bonus to reflect the relative contributions to savings and revenue generation made by individuals and teams. Based on the past statements of profit and loss, you can calculate a ***specific and measurable*** profit level that will be ***realistic, timely and attainable*** by your team.

Peer recognition is important to the functioning of a team. A good salary and incentive are only two forms of recognition. One of the most effective ways to motivate a team member is by making him feel important as a contributor to his team, his project and his company. The utmost thing that you can to do is getting the team to buy into the goals, incentives and recognition.

15.3. Working with the Team

A CEO is empowered to design the organizational structure, assemble the team and define the working relationships among the various entities. In doing so, the CEO aims to:

- Ensure that important work gets done efficiently
- Enhance the company's performance
- Aid delegation of job and authority
- Facilitate communication and collaboration
- Encourage team work
- Provide transparency of team management, operation of the company and awarding of incentives.

In the Assessing Phase of your company's development, most of the company work will involve the feasibility and design of the new invention. The Launching Phase involves more work on manufacturing and prototyping. During the Building-up Phase, you expand your capabilities to include production, servicing and marketing. To perform these tasks well and in a short period of time, you need to form a cross-functional team that is staffed with experts in:

- Marketing
- Design engineering
- Manufacturing engineering
- Service support
- Production engineering

- Accounting and inventory.

All aspects related to the functions above are done concurrently, especially during the Building-up Phase, in order to reach the stage in which revenue and profit will be generated.

If the team is to succeed, the CEO must become proficient in leading, and simultaneously participating in, the engineering team. In getting things done, the team leader needs to pay attention to team discipline, team learning, intra-team relationships, conflict resolution, and other factors that affect the team's performance.

All members need to know the team's objective, although each member will work on those things in which he or she has expertise. The efforts of all are integrated to produce a product for the market that achieves any or all of the following:

- Reduction of product development time
- Reduction of the number of engineering changes
- Effective outsourcing some manufacturing and production work
- Innovation and improvement of a product
- Reduction of manufacturing costs
- Improvement of product quality, reliability and serviceability
- Reduction of the time-to-market.

This engineering team consists of a small number of persons (or all persons people in the small business venture) of complementary expertise and who are committed to work together to achieve the above impacts.

Rolling up your sleeves and willingly do whatever is necessary are essential in being the team leader. Never think of yourself as being too good for a job. In order to encourage team members to learn better and faster, motivate them to learn new technologies, new thinking and new marketing skill. Outsourcing work may be the norm for a small company, if cost effective and appropriate. If you have transparency in decision-making, delegation of jobs and authority, individual accountability, and incentives, you will have a cohesive and enthusiastic team to move the product development toward the marketing phase.

The management of people and project will be a challenge as business development grows, activities expand beyond the company, and team members are dispersed. Web-based tools for project management are available to assist you in dealing with the following issues:

- Task and resource management
- Cost management
- Risk management
- Communication
- Knowledge and documentation management.

The ideas covered in this subsection were drawn from two chapters of the engineering management book by C. M. Chang (2). These chapters should be consulted for more ways of "Organizing" and "Managing" your team and company.

15.4. Building the Winning Team

To build a winning team for the company, your first task is to recruit people with appropriate expertise to work as team members and who can also honor the set of rules set for the team. Here are some working rules that have been extracted from the book by Blair Singer (3); you should consider adopting for your company and getting the team to agree to abide by:

- *Be willing to stand behind the purpose, rules and goals of the company.*
- *Be willing to do whatever it takes to support any and all team members.*
- *Have a willingness to stay together.*
- *Support early, often and unconditionally.*
- *Speak supportively and with good purpose.*
- *Complete your responsibilities and assignment.*
- *Make only agreements that you are willing and intend to keep.*
- *If a problem arises, first look to the system for correction and then communicate your solution to the person who can do something about it.*
- *Do not go behind a person's back with a problem.*
- *Agree to work toward an agreement.*
- *Accept personal responsibility. No laying blame, justification or finger pointing.*
- *Do not let personal issues stand in the way of your work.*
- *Make a commitment to win and allow others to win.*
- *Clarify your own communication and verify the response.*

The second task is for you and the team to enforce the rules. Before getting to the point where you show the misfit, troublemaker or non-team player the door, Singer suggests that you just "CALL IT" using the following tips:

- *Pick an appropriate time to make the call, but don't wait too long.*
- *Acknowledge how you are feeling first to the other party*
- *Get permission from him or her to make the call.*
- *Correct the behavior, not the person. Let the rules be the policeman.*
- *State specifically what does not work and offer support.*
- *Make clear what the benefits of correction are for the team and for the individual involved.*
- *Thank the individual for listening and listen to his or her response without interruption.*
- *Acknowledge the individual's proper behavior later when you see him demonstrate it.*

Use the following tips to facilitate the team operation:

- Be concerned about over-management. It may be a waste, is often a hindrance to progress, and, worse yet, a morale killer. Reduce the number of management layers to an absolute minimum.
- Encourage debate on understanding, thoughts and concerns before a decision is made.
- Discourage nay saying, such as "I told you so," once the decision is made or if something does not go as planned.
- Instill in all team members the importance of documentation. Good documentation will not only assure a good understanding of what has been done, but also be important in substantiating the validity of a patent, assuring legal protection, and facilitating job continuation and/or reassignment.
- Be prepared with a plan B for critical tasks. No matter how well you plan, problems will always arise. A well thought-out plan B will assure an orderly progress of tasks.
- Require some managers to do technological work to keep themselves up-to-date on technology advancement.

Finally, as a team leader, you need to work on the ability to:

- *Spot and leverage the strengths of others.*
- *Teach others how to succeed.*
- *Use mistakes to strengthen and grow the team.*
- *Use frequency of interaction to build relationship, consistency and, most of all, trust.*
- *Promote a realistic, but brighter, future to the team.*
- *Sell a vision, an attitude, the rules and ideas to the team.*

Technology companies should examine a few exceptions to the teamwork requirements. As stated by Bakken in Reflections on Leadership (1): "*Much success of Medtronic has been due to brilliant mavericks that were difficult for our engineers to work with. Because of the lead of mavericks, Medtronic effected early breakthroughs (in implantable pacemakers). With an enormous amount of imagination and patience, the most effective leaders accommodate the ornery thinkers and give them the attentive ear that their ideas deserve. The challenge to leadership is great, but so, too, are the potential rewards.*"

References

1. Bakken, E.E., *Reflections on Leadership*, Medtronic, Inc. (2001)
2. Chang, C.M., *Engineering Management, Challenges in the New Millennium*, Pearson/Prentice Hall, NJ, pp. 44-72 and 373-428 (2005)
3. Singer, B., *The ABC's of Building a Business Teams that Wins*, Warner Business Books, NY, 2004.
4. www.topachievement.com/smart.html

Chapter 16

Manufacturing Your Product

16.1. Procurement and Outsourcing

Small biomedical companies often start as a one- or two-product company. Its initial expenditures will be the one-time cost of designing, manufacturing and testing a new product, as described in Fig. 16.1, before the company can proceed to scaled-up production. The design

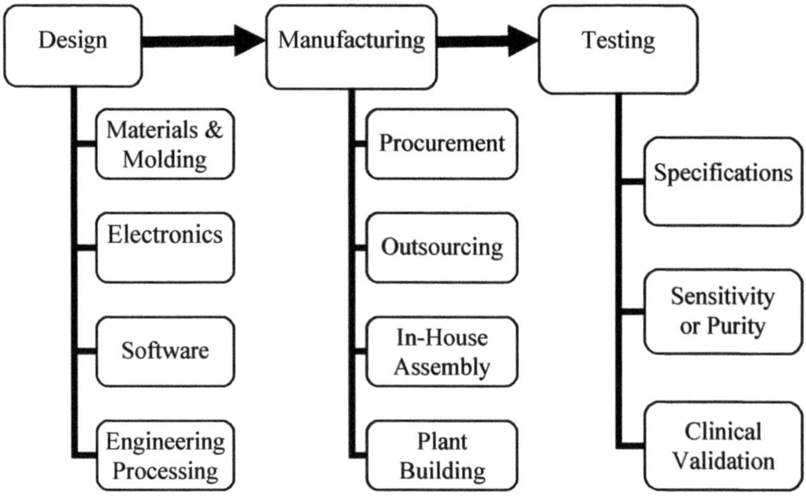

Fig. 16.1. Record keeping, as required by Good Manufacturing Practices of the FDA, for processing blocks that might be involved in the manufacture of your medical devices or pharmaceutical products.

step may include the selection of materials, design of electronic circuitries, writing of software, and/or selection of engineering processes. As the company approaches the manufacture of the first few prototypes, it will become involved in procuring components, finding an outsourcing firm, performing in-house assembly, and/or making plans to expand plant facilities. After test results show that the product matches the specifications, has sufficient sensitivity or purity and demonstrates effectiveness and safety in clinical usage, the company will apply for product approval from the FDA. Only then, can the company scale up the production and move to the marketing phase.

Several pointers on how best to procure the components of your medical inventions or outsource the design, testing and manufacturing of them are described in this section.

There are at present many companies in the USA and the world that have the expertise necessary to carry out this non-recurring engineering for medical device and pharmaceutical companies.

Procurement. Suppose that your new invention is so simple that it consists only of three components. Each component may only require a few modifications to a product that is already available in the market. You contract several suppliers to remake the modified components, and then you assemble the components into the product for marketing. In order to protect your intellectual property (IP), you can subcontract the manufacturing of the three components to three different suppliers. If your competitors have no instruction manual that explains how the new invention operates or any knowledge of how these components will be put together, your IP right may be somewhat protected for a while.

Even with this advantage, you should still thoroughly conduct due diligence on each supplier before you sign a purchase contract that clearly spells out the specifications of the component to be manufactured, the delivery date, the shipping cost, and the terms of payment. The second due diligence is conducted to obtain references that reveal how reliable this contractor is in delivering the product as specified. Third due diligence is to assure that the manufacturing practice of the contractor follows FDA GMPs or ISO Standards that are appropriate for your company's medical device or pharmaceutical product. The fourth due diligence is intended to assure you that the contractor will fully transfer his/her Good Manufacturing Practice (GMP) records to your company after he has delivered the goods.

Once you receive the goods, you should immediately identify the batch and check the products individually or randomly to be sure that the specifications have been met. Call the supplier right away if the delivered product does not meet the specifications. The names of the inspectors who checked the product specification should be entered in your records. The key message here is that you must have the records necessary to demonstrate that diligent and verifiable efforts have been made in the manufacture and assembly of the medical product.

Outsourcing. When your product begins to become more complex, it may be more practical for you to outsource the design, testing and manufacturing to a firm that specializes in your innovation's field and has done completed jobs similar to yours.

Use your connection from previous employment to identify such a firm. Your friends working in medical device or pharmaceutical companies may give you name recommendations. There are also a number of websites that you use to identify outsourcers. It is important to be aware that the collaboration between your company and the outsourced firm will be as intense as a partnership (4). Both parties need to do due diligence to assure themselves that they will have a working partnership to manufacture the product properly.

The product development may encounter some unexpected events. Both partners must possess the motivation to work out any problems.

An outsourcing agreement will be somewhat like the licensing agreement that was described in Section 14.4, but will include a description of the development, design and manufacturing work to be done by both partners. On your part, you will need to ensure that the firm to which you outsource the manufacturing is appropriate, capable of doing the work and has the right people to focus on the design and manufacturing work and the facilities in which to carry it out. In addition, the firm should show that its manufacturing follows the FDA good manufacturing practice and be able to sustain the services required for the continuous improvement, development and manufacturing of your devices after the conclusion of the current agreement.

An Illustrative Example. You can divide the outsourcing work into several phases as shown in the following case involving the development of a 5 MHz ultrasound processing system. This system will use a newly produced, dual channel, A/D converter and a new algorithm to determine if the use of ultrasound has enough sensitivity to measure some physical characteristics of biological fluids. Because of the A/D converter's low price, the task will be to examine whether the A/D converter that has a sampling rate of 60 mega samples per second and digitized data at 16 bits can really achieve better accuracy than that offered by a competitor's system, while achieving a reduction in manufacturing cost.

To test the feasibility of this proposition, you prepare a contract for the contracted firm to use the evaluation board of the A/D converter to digitize and store the ultrasound signals and then use a PC to perform the computation.

If sufficient accuracy is demonstrated, you ask the contractor to design a full circuit board in the second phase that will integrate the A/D converter and a microprocessor into a prototype for you to evaluate its performance. The outcomes of this phase will be a bill of materials (the listing of all chips, resistors and capacitors used in the electronics), the schematics in Gerber files, and the software.

At this stage the assembly of chips onto the circuit board of the prototype will be largely done by hand. It is also essential that you invite a third party to validate chip by chip the bill of materials and schematics provided by the contracted firm. It is much easier to find errors and misrepresentations in documents or products at this stage than later.

The third phase will be to demonstrate that the chips can be assembled in large quantities by machines with the same circuitry performance. In this way, you will more accurately learn the manufacturing cost of the ultrasound system and the assurance of quality products when they are manufactured in large quantity.

This development process is complicated. The first outsourcing of design and manufacturing will be a learning experience. On many occasions, face to face discussion and direct communication are so essential in leading to better solution. Major changes can be made more easily by direct communication. Try to find a firm nearby to work with you in the development of your device.

After gaining more experience and knowledge about outsourcing, you can use the internet to investigate the possibility of outsourcing your manufacturing to firms in other regions of the US or overseas. In fact, it is likely that the firm with whom you have contracted the manufacture of your product already has a subsidiary or partner in China or other countries that is accustomed to mass producing quality products.

If you outsource internationally, the manufacturing facilities that do your work must practice the same GMP as the US manufacturer. Because of the large quantities being ordered, a foreign company will set up an assembly line to ensure that it can assemble your product at lower cost and a lower failure rate.

Because of rapid market growth and the emergence of new industry-spanning market segments, medical device manufacturers have begun to seek partners that can provide full service outsourcing, completing all phases of device development from market research to prototype building, equipment design and manufacturing (2).

Advantages of Outsourcing. Here are some of the ways in which your company may benefit from outsourcing (1).

- **Multidisciplinary and Experienced Team.** You are in the position to choose the firm to which you will outsource your work. You can check out what it has accomplished and whether it has the expertise and experience necessary to do your work. In the preceding example, expertise in ultrasound, hardware and software, and more specifically on high frequency signal processing and signal digitization without interference and mis-synchronization, was the key to successful completion of the project. It may be difficult for you to build up an in-house team to do your one-time project. On the other hand, this is the type of job being done regularly by many outsource firms. Consequently, your subcontractor may have already lined up the team to work on other projects after they have completed yours.
- **Expanded Engineering Capability.** Just like having a design and engineering team, the subcontractor may have the necessary engineering facilities to do work that is not practical for you without building an appropriate facility in house for a one-time use.
- **Fewer Errors and a Reduction in Development Time.** The expertise and experience of the outsourced firm should result in fewer errors and enable the development to be completed sooner.
- **Commitment in IP Protection and Confidentiality.** The design and engineering firm knows the importance of protecting your IP rights and keeping your project confidential. Most likely their administration already has a policy in place to assure compliance by the outsourced firm and its employees.
- **Better Record Keeping and Compliance with Regulations.** Ensure that the firm has done the design and engineering work on medical device or pharmaceutical product of the same field. Specify in the engagement agreement what records must be conveyed when the work has been completed.

Outsourcing Precautions. The success of the partnership depends greatly on the relationship between your company and the outsourced firm. Here is why the relationship is important:

- **Commitments.** Because you are the inventor you are the only person who will know whether the manufactured device can have the functions that you wish. The design and engineering firm will try its best to accomplish what you want. Sometimes a product may not have all the specified functions due to limitations in electronics, chemistry or physics. You will remember that the portable, battery-operated pacemaker mentioned previously could be developed only after the arrival of transistors. The contracted team may recommend changes in the design and engineering of your product, but you are still the one who must make the final decision. You work on the premise of trust in your outsourcer. However, trust yourself even more as you must pay the bill, even if the team does their job according to the contract, but fails to produce a product that meets the specifications.
- **Communications.** It is a good practice to maintain good communications as the project progresses. Schedule periodic meetings of both parties to evaluate the project's status and to make necessary changes and modifications to the design.
- **Expected Costs and Timelines.** The project budget and timelines should be set at contract signing. Appropriate resources and manpower should be allocated. You must be prepared for unexpected events in product development, whether the work is done in-house or by outsourcing.
- **Dispute Resolution.** Try to develop a good partnership with the least chance for it to go wrong. However, be realistic so that disputes over work and payments can be resolved.

16.2. Current Good Manufacturing Practice (cGMP), Accountability and Risk Management

cGMP. GMP is a set of FDA regulations developed to promulgate Good Manufacturing Practices (5). These regulations require that manufacturers of pharmaceutics, medical devices and food products take proactive steps to ensure that the products are safe and effective. The regulations are designed to assure the quality of manufacturing and minimize or eliminate contamination and errors. The regulations also stipulates what is required of manufacturers to meet in matters of record keeping, personnel qualifications, sanitation, cleanliness, equipment verification, process validation, and complaints.

cGMP are open-ended regulations that do allow individual manufacturers to decide how best to implement the controls or to interpret the regulations in a way that is favorable to that company or industry. The "c" preceding GMP stands for "current," meaning that manufacturers must employ the most up-to-date technologies and systems to demonstrate their compliance with the regulations.

The guideline on **Pharmaceutical cGMPs for the 21st Century: A Risk-Based Approach** states "The management is responsible for establishing the quality system structure appropriate for the specific organization. Management has the ultimate responsibility to provide the leadership needed for the successful functioning of a quality system." It is essential that all company personnel practice GMP individually by complying with the appropriate standards established for the particular type or class of employees.

A Quality Assurance Department should be established to set the standards and assure their compliance. This needs to be expanded to include compliance by suppliers and subcontractors. It also may be the best if you arrange to have an external consultant check your compliance. If you are notified by FDA that your facilities will be inspected on a certain date, get your records, facilities, and presentation ready to show to FDA staff when they visit your company.

Accountability. The following consent decree may describe the trend toward increased accountability to the quality assurance function of pharmaceutical, medical device and biotechnology companies. In January 2009, the FDA released the news that it had filed a Consent Decree and was awaiting the court's entry of a permanent injunction to bar not only a generic drug manufacturer, but also the CEO, VP of Corporate Quality and a board member from manufacturing and distributing 20 or so drugs until the company and its officers were in compliance with the cGMP. The FDA had ruled the company and its officers guilty of manufacturing and distributing unapproved drugs within the United States.

Because of this consent decree, these officers were held additionally accountable when the company's shareholders sued for security frauds on the grounds that the officers had misled investors about problems in the company's manufacturing facilities and had failed to bring the facilities into compliance with federal regulations, including those of the FDA.

A review of the FDA's records also indicate an increasing number of warning letters being issued to the executives of life science companies for issues related to manufacturing compliance.

The pressure on pharmaceutical, medical device and biotech companies to comply with FDA regulations certainly will have a significant impact on the operation, business strategy and resource allocations of your company. As its chief executive, you must devote more effort to assure that your firm's quality oversight and compliance activities have been documented and demonstrated to be effective to both regulatory agencies and your shareholders.

Risk Management. Even with the best compliance, unexpected incidents with adverse consequences can still happen. You need to have a response plan to inform internal personnel, FDA and the public of any adverse incident and to decide on a series of actions to identify the cause of the incident and the need for recall. As A. R. Lentiani, Jr. advised, you need to be conscientious and avoid the following five mistakes (3):

- Having no response plan or internal notification system
- Putting lawyers in charge
- Shutting out the news media
- Using untrained spokespersons
- Lacking an outside perspective.

To survive such a crisis, Lentiani stated that:

"Bad things can happen to good organizations. The trick is to respond immediately and proactively in a direct and honest manner. That sounds easy, but when you are up to your neck in alligators, you might not be thinking clearly. That is why you need to have a plan, test it regularly and follow it.

The other important thing to remember after any crisis is to investigate the causes thoroughly and then correct what caused the problem in the first place. The more serious the crisis, the more important it is for you to determine the contributing factors and fix them so a similar event never happens again."

16.3. Lifecycle Management for Maximum Value

The implementation of cGMP is the best way to assure the effectiveness and safety of patients who use your product. You will have to invest manpower to enter data and to analyze the collected data. The added cost will be significantly outweighed by what you can gain if you provide quality service for patients. There are many data management programs on the market that you can adopt for your operation. Several additional benefits are highlighted below. The last three cover the management of the total lifecycle of the product described in Fig. 16.2.

Better Quality Assurance. Fig. 16.2 illustrates the manufacturing lifecycle as one progresses from design to small scale manufacturing, testing and finally large scale manufacturing. The GMP is a reliable

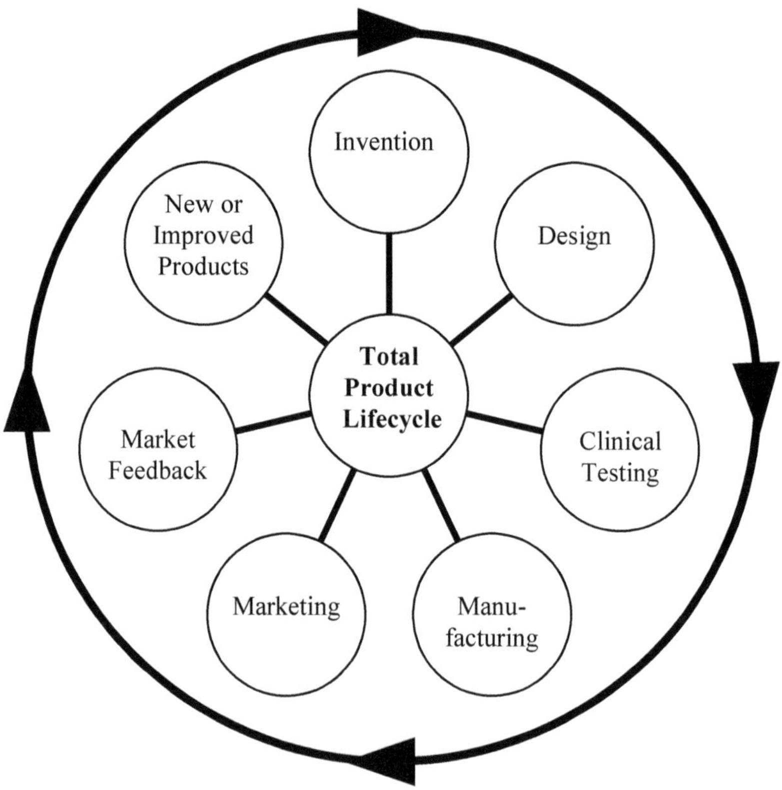

Fig. 16.2. Data Management during the total lifecycle of the product from invention to new improved product.

means to achieve quality of your product. If one or more processes during the lifecycle were not undertaken in accordance with the GMP, identifying the irregularities will enable the operator to stop and not continue to the next process.

An on-line quality monitoring of the manufacturing process is much better than an off-line testing procedure. First, an off-line testing may rely on random samplings, which may mean that only a small subset of the products will be tested. Secondly, identifying irregularities can only be revealed by an analysis of the final products. Thirdly, standardized and widely accepted GMPs that are implemented in life management serve as a good vehicle by which to assure quality consistency.

More efficient operation. A hypothetical case that can occur frequently in the design of chemical processing facilities, is used below to illustrate why the implementation of lifecycle management can lead to more efficiency in operation. A pharmaceutical company wanted to double its production capacity by building another, identical plant nearby. The projected construction cost was $12 million. Using data in the records that the GMP required the company to maintain, the consultant, who was hired to evaluate the feasibility of plant expansion, was able to identify a few bottlenecks in the processing plant. As a result, resolving those bottleneck locations enabled the company to double its production at a reconstruction cost of about 10% of the projected budget to construct the second plant.

Similarly, you can model the dynamics of the manufacturing process to determine how to reduce inventories, improve risk management (e.g., financially or by reducing material delivery delays), and to reduce manpower costs. With the globalization of trade, there is a possibility that, for example, a snow storm in China may delay the arrival of your product by, say, two months. Be conservative and play it safe in the modeling analysis.

Improvement in Software Safety. It is so easy to revise your software program to enhance the functioning of your medical device. However the FDA increasingly has been finding serious problems with device software because of aging microprocessors and platforms and the addition of new functions and shortcuts in the software. It is also possible that a new evaluation procedure may not be compatible with older microprocessors and platforms for problems identification. Software problems can cause malfunctions in device that could lead to product recall and, in the worst instances, to the death of patients.

With good record keeping for design and software development, you may be in a better position to select the right diagnostic program to verify and validate your software, to prevent defects and failures, to conduct risk analyses, and to take corrective actions or upgrade your hardware.

Shorter Reaction Time. Suppose that there is a fire in a processing plant. It may be simply human nature that causes the plant manager to postpone reporting the incident to top management until some other key problem has been resolved. However, it is the CEO who has the ultimate responsibility for ensuring that the problem is addressed correctly and promptly. The on-line entry of records and timely alert of top management by a record management program can get more people involved in generating a better response in a shorter time to correct problems or incidents of an adverse nature.

Facilitation of New Product Improvement. Integrating marketing intelligence and patients' evaluations in a total lifecycle data management system can result in further product improvement and development at a faster pace and with more efficiency.

Moral Enhancement. Suppose your sales rep just landed a big contract. He entered the sales into the record and described how he made the sale. Would you like to be alerted by the record management program instantaneously so that you could call to congratulate the sale rep personally well before the sales rep advises the supervisor who then calls the VP Sales before the message works its way to you!

References

1. Andrews, R. R., MD&DI's Guide to Outsourcing, S32-S43, June 2006.
2. Andrews, R. R., MD&DI's Guide to Outsourcing, S38-S48, August 2006.
3. Lentini, A. R., *Risk Management Magazine,* June 2009.
4. Thompson, H., MD&DI's Guide to Outsourcing, pp. S52-S61, June 2006.
5. www.gmp1st.com/gmp.htm

Chapter 17

Marketing and Sales

Each year, the FDA approves 3,000 to 4,000 devices under the 510(k) program. Because its approvals are based on showing of substantial equivalence, many of these devices will represent incremental advancements from their competitors' products. Medtronic stated in its 2007 Annual Report that two-thirds of its revenue came from products introduced in the last two years (5). The problem of competitions may be further compounded by the fact that your company has only limited funds for marketing purposes.

In comparison, only 30 to 60 new drugs per year are approved by FDA through its PMA program. For pharmaceutics and biotech companies, the drug development attrition rate comes to the yielding of 2~3 drugs out of 100 projects. In recent years the estimate on the total expenditure for the development of one successful drug can be as high as $1 billion.

For such a challenging market, evidence-based information, device designs centric to the patients and well-differentiated products are essential to your sale representatives if they are to reach your customers. Then your sales presentation must exhibit sufficient understanding to win the customers and obtain their purchases. We provide some background here to help you become more knowledgeable about your customers, market characteristics that are important in marketing the medical device or product to the health or pharmaceutical industry, customer relationship management (CRM) and marketing ethics. This should help you prepare your marketing plan. This chapter concludes with a list of points for your use in developing a marketing plan that is suited to your medical device or pharmaceutical product.

17.1. Know Your Customers

Traditionally, customers who have made a decision to acquire your medical device or pharmaceutical product are prescribers, primarily physicians. As our healthcare costs rise from 5% of Gross Domestic Product (GDP) in 1965 to 18% in 2015, the three categories of customers shown in Fig.17.1 are those who have significant influence in the decision to acquire your product. The first category is represented by patients, nurses, clinical engineers, and physicians who use your product. The persons in the second category prescribe or purchase your product. They include physicians and the value analysis teams of hospitals and managed care organizations. The third category consists of the payers and policy makers. These two are grouped together because, in many occasions, they are parts of the government or an insurance company. The following text discusses these groups of persons for your consideration in the development of your marketing and sales plan.

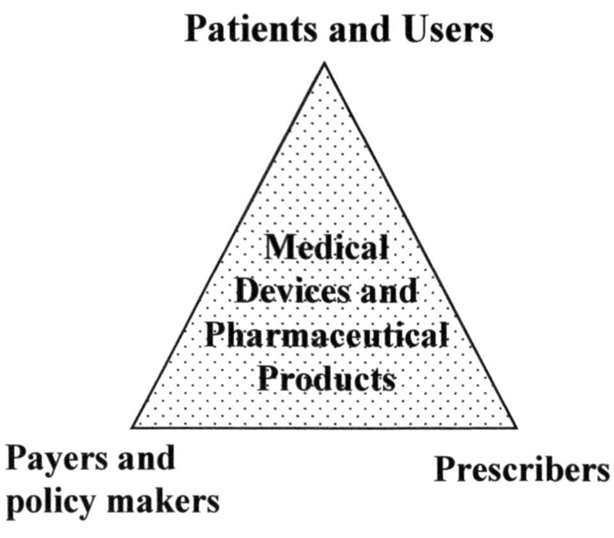

Fig. 17.1. Three categories of persons who are involved in the decision to purchase your product.

Patients and Users. It is up to you to convince patients, physicians and users that your medical product can improve patient experience and outcomes, maximize compliance and reduce risks. The pharmaceutical and healthcare market research company Manhattan Research (9) advises that:

- More than 60% of U.S. adults turn to the Internet for healthcare and disease information.
- Physician interest in live video detail made available by sale representatives is growing. Currently 45,000 US physicians meet with sales representatives via online video and more than 300,000 more show interest in interacting with sales or other company representatives online.
- The increasing consumer reliance on the Internet for health and pharmaceutical information is transforming digital direct-to-consumer (DTC) marketing from a "nice to have" tactic to an essential communication channel for companies to use to reach customers.

In view of the great number of devices and products on the market, there is less time for physicians to meet with your sales representatives. In essence, your marketing effort must take advantage of an effective means of digital communication to convince physicians to purchase your medical devices and pharmaceutical products without actually meeting your representative.

Equally important will be patient-centered Internet communications and webcasts that serve to educate patients about their own healthcare management, empower them to work on improving their health outcomes and build consumer trust in and loyalty for your company's technology.

Physicians, nurses, clinical engineers, and pharmacists can be users of your medical devices in their work. They want to use devices of high quality and backed by good service. Their opinions are solicited by the hospitals when making the purchasing decision. Good quality and service should not only be your company's goal, but also an important message that is embedded in your marketing and sales effort.

Prescribers. The marketing objective relating to prescribers is to increase the number of times that your product is prescribed. Hospitals and managed care organizations are becoming an important segment in the purchase of medical products as physicians are no longer the only ones who are involved in the purchasing decision. Operating with a low margin, hospitals and managed care organizations are under tremendous pressure to maintain their financial integrity. 25% of hospitals in the US have negative total margins. The future financial prospects of hospitals are further constrained by:

- Weak growth in numbers of patients.
- A lower reimbursement of health-related expenses from insurance companies and governments.
- A decrease in the collection of fees.
- An increase in costs of operations.
- The impact of the nation's financial crisis.

Hospitals are tackling these problems on many fronts. In the 1990's hospitals were using group purchasing organizations as a means to contain costs. To counter this trend in bargaining power and to facilitate the compliance of government healthcare policies, many medical device companies merged or banded together to form larger organizations. In today's hospitals, the following three cost-containment strategies are particularly worthy of consideration by your marketing and sales team:

- ***Vendor Access Restriction.*** Physical access by sales representatives to hospital physicians and staffs can lead to the increased use of the products of those sales representatives. As a means to reduce the influence of the sales representatives, guidelines have been established to limit sales representatives' access to staffs and physicians.
- ***Data Driven Supply Chains***. The measure of procurement success by the hospitals is changing from the reduction of the cost of acquisition and operation per unit to effective patient outcomes at lowest possible cost. As a result, hospitals are taking a more active role in managing the sampling and trial of medical products. They require that sales representatives have credentials and prohibit the promotion of unapproved products.
- ***Value Analysis Teams.*** With the advances in information technology, hospitals are more knowledgeable today about their inventories and usage and have more access to the data required to reduce the expense of supplies and operations. Policies on costs and effectiveness have empowered value analysis teams to make purchasing decision. The teams include physicians, hospital administrators, people employed in materials management and finance, nurses, pharmacists and clinical engineers. Value analysis with evidence-based medicine and

government requirements concerning quality improvement of healthcare will continue to play important roles in the process of product evaluation and acquisition.

Payers and Policy Makers. Your objective with this category of persons is to establish a funding policy and decisions that are favorable to your product and to create or revise healthcare policies in order to build the demand for your medical devices and pharmaceutical products.

Health insurance is insurance that is provided by a government-sponsored social insurance program or by a private insurance company that has a contract with an individual or his/her employer. The employer or individual pays premiums or taxes to the insurance company in exchange for its assistance in protecting the individual from high or unexpected healthcare expenses. The insurance company will pay healthcare providers "reasonable and customary" fees for the services provided to the individual. If the providers are in the "Network," they will accept a discounted rate from the insurer. The benefits of an insurance plan and the rules about co-payments and out-of-pocket maximums are specified in the contract. One subject that is important for a medical product company is whether the insurers have established "reasonable and customary" charges for the use of your medical devices or pharmaceutical products.

The largest insurer in the US is **Centers for Medicare and Medicaid Services (CMS)** of the Department of Health and Human Services. This organization operates three Federal healthcare programs: Medicare, Medicaid, and Children's Health Insurance Program. Its mission is to ensure effective, up-to-date healthcare coverage and to promote quality care for its beneficiaries. The CMS 2008 Fiscal Year budget was $458 billion.

To learn more about CMS programs and payments, we will review, as an example, its hemodialysis operation (10). The 2006 Medicare claims data indicated that there were about 4,700 clinics that furnished outpatient maintenance dialysis to an estimated 315,000 Medicare dialysis patients. A total of $8.1 billion has been paid to date by Medicare for expenditures on dialysis and related drugs totaled - or $25,700 per person. The average Medicare payment per dialysis treatment in 2005 was $237. Of this amount, $143 was for partially bundled prospective rate covering dialysis and $94 was for related drugs and laboratory services.

If you are a Medicare patient who is receiving care, you will receive a Medicare Summary Notice like that shown in Exhibit 17.1. The five-digit number in the first column of each line is the code number in the Healthcare Common Procedure Coding System (HCPCS), for that sub-classification, based on the Current Procedure Terminology (CPT) codes. The CPT was developed and maintained by the American Medical Association (AMA) to provide identifying codes that accurately describe medical, surgical and diagnostic services and procedures that healthcare providers use. The second to the fifth column of the exhibit show the amounts charged by the service provider, what was approved by CMS, how much will be paid by CMS to the provider and the amount that may be charged to your supplemental insurer or yourself. There are deductibles that apply when determining how much you will need to pay for all of medical services and drugs that you receive in a particular year.

Exhibit 17.1. A typical billing notice from CMS for a Medicare patient

CMS				Medicare Summary Notice
Services Provided	Amount Charged	Medicare Approved	Medicare Paid Provider	You may be billed
Emergency dept visit (99285)	$492.00	$161.79	$70.35	$91.44
Electrocardiogram report (93010)	$46.00	$8.41	$6.73	$1.68
Office/outpatient visit, new (99203)	$188	$92.48	$73.98	$18.50
Office/outpatient visit, Est (99213)	$122.00	$61.15	$48.92	$12.23

If a device is already used in services and has HCPCS codings and yours can serve as a replacement, the provider will be able to compare the profit margin it would have by performing the procedure with your medical device instead of the device it presently uses, and then decide whether to replace the presently-used device with your device for future procedures.

If you have invented a device for use in a new procedure, you should first request that AMA grant you a CPT code for the procedure. Then, you should request CMS to evaluate this new procedure and to establish the amount that Medicare will approve and pay to the provider. The creation of a new code and introduction of a new item for reimbursement is time consuming and expensive.

Then, private insurers may follow by establishing their own reasonable and customary rates for the use of the procedure. Only if approvals for the coding and reimbursement of your device have been obtained will the providers consider purchasing it.

For FDA to approve your device, the effectiveness and safety of your device are of utmost importance. However, the medical need for the device is the first consideration of CMS when setting the reimbursement rate.

In the case of hemodialysis care, the Medicare Modernization Act requires the Secretary of the Department of Health and Human Services to issue a report to Congress that describes in detail the establishment of a fully bundled Prospective Payment System for patients with End Stage Renal Disease by 2011. The System to establish the reimbursement for hemodialysis will account for such factors as the current base rate of $235 per treatment, a facility level adjustment for the wage index, and patient-level adjustments for age, gender, body surface area, low body-mass index, and duration of replacement therapy. The chief Medical Officer of CMS emphasizes that a "quality incentive program" will accompany the new payment structure (10). "*Instead of paying only on the basis of quantity*," he has said, "*Medicare will also begin rewarding dialysis providers on the basis of quality standards, such as how well they manage patients' side effects.*"

Payments for dialysis come from three sources (8):

1. Medicare (the Federal insurance program). CMS sets the reimbursement rate and pays about 80% of that rate to dialysis clinics for 75% of US dialysis patients.
2. Medicaid (insurance available through states for those who have limited incomes and assets). About 10% of all dialysis patients have only Medicaid as insurance.
3. Private insurance, such as an employer group health plan (EGHP). About 10% of dialysis patients have private insurance. Clinics charge EGHP two to ten times the Medicare-allowed rates. What private insurance pays is a closely-held secret, but the reimbursement, on average, would be twice Medicare's rate.

Under many current healthcare plans, providers are paid for each service performed. Thus, providers have a strong financial motivation to perform as many services as possible. The concerns about malpractice may lead to over-prescribing and over-utilizing healthcare services.

Health Technology Assessments (HTA) now constitute the driving force of changes in the delivery of healthcare (6). Regulatory agencies, government and private sector payers, managed care organizations, hospitals and healthcare networks are some of the organizations that undertake HTA. Regulatory agencies, such as the FDA, want information and/or advice to help it decide whether to permit the commercial use of a particular drug, device or technology. Healthcare payers want to know whether technologies should be included in health benefits plans or disease management programs. They use HTA to address such issues as coverage and reimbursement for use of medical devices and pharmaceutical products. Clinicians and patients want to know whether it is appropriate to use certain technologies for the particular clinical needs of patients. Hospitals, healthcare networks and group purchasing organizations use HTA to decide whether to acquire and manage certain technologies. Your company and investors can use HTA to help decide product development, marketing, venture funding, acquisitions and divestiture issues.

In recent years, "***Value Based Purchasing***," also known as "***Pay-for-Performance (P4P)***," is being implemented by many private insurers and CMS to improve the quality of healthcare and to slow the rate of growth of healthcare costs by linking compensation to providers to their certain performance measures (11). California now has the largest P4P program in the country (2). However, there are not sufficient data to fully determine whether P4P programs, that were just adopted, do improve healthcare and reduce costs. For P4P programs, you should give particular attention to the subject of how your product will influence the performance measures that are used to quantify the improved care for patients.

In summary, if your product is an improvement of an existing product, healthcare providers can use the approved fee to pay for the use of your product. For a product that provides a new service, you should incorporate your knowledge of CMS, HTA and P4P into a system to gather evidence-based information for you to gain coverage from private and public healthcare payers.

17.2. Market Characteristics of Medical Devices and Pharmaceutical Products

The wide spectrum of customers and intermediaries has disrupted the traditional sales relationship between sales representatives of medical device companies on one side and patients and physicians on the other side.

Education. Education of users is still an important way to get physicians to know about the product as you introduce it to the market. Courses, workshops and conferences to publicize products are becoming less popular with physicians because of federal regulation of marketing ethics and the time and cost to them in being away from their offices. With the growth in the use of internet, on-site activities are being replaced by online institutes, enabling physician to attend lectures, seminars, demonstrations of products and procedures and receive comprehensive training and follow educational programs without leaving their offices. Digital marketing is a trend that medical device and pharmaceutical companies can use to leverage internet technology to improve their relationships and interactions with customers (9).

Patients can exert significant influence on the purchase decision, as evidenced by the extents we see in patient-targeted television and magazine advertisements. Of course, the influence of a patient would also depend on whether his or her insurance pays for the purchase or use of your product. Communication of medical product companies with patients on the web and through digital marketing is actively pursued. They set up websites to help patients understand the causes of their medical problems, treatment options, the product's pros and cons, and information about health and disease recovery with use of the product.

Product Lifecycle. The lifecycle of your product has three phases: Introduction, Growth, and Maturity and New Products (Fig. 17.2). In the Introduction phase, the product must "support itself" to varying degrees with increasing numbers of sale successes and by painstakingly building a reputation for quality and trustworthiness. Evidence-based information should be available for your customers to assist in their purchase decision. Resources are allocated for you to carry out promotion and training activities, such as on-line symposia and workshops for education, advertising in magazines and televisions, and training of the sales force.

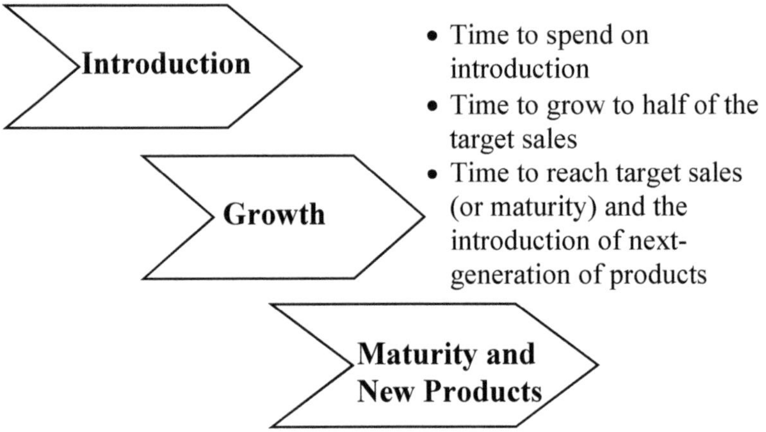

Fig. 17.2. The product life cycle. It is described by three times that appear in the right upper corner.

As your sales increase to a much higher level during the Growth Phase, you should invest further because technology in the field moves quickly, a product life cycle can be short, and even a successful product can be overtaken by your competitor. More real life evidence should be collected during the Introduction and Growth phase to demonstrate the cost/effectiveness of your product to customers. During the Growth and Maturity phases, your company can introduce a more powerful product to help your company continue its dominant position in the market

Healthcare needs are not driven solely by economy. The demand for medical devices or pharmaceutical products is based on personal health, insurance policies and government rules and funding. The key is to sell your customers on the benefits of your product that differentiates it from those of your competitors. Marketing involves communication with your customers. You need to ensure maximum exposure for your new product by making each sale representative not only a spokesperson for your company, but also a specialist in your product's field. Your websites need to be interesting to patients and physicians, especially the explanation or description of how your product differs from those of your competitors. Only then will your selling message gain credibility.

Pricing. Your product's price may be based on the cost to manufacture your product, the manufacturing overhead, and your patent position. The economics of your product to hospitals and managed care organizations needs to be considered. If price is one of your advantages, that is great. However, remember that you need service and quality to keep your competitors away and to secure your customer relationships for the future. Evidence of the cost/effectiveness of your product should be considered when setting the price of your product.

The European market has been moving toward greater evidence-based pricing for medical devices and pharmaceutical products. Experts in the US are calling for the use of comparative research of effectiveness to help policy makers determine whether certain therapies and tests are reasonable or necessary. However, there is one unique aspect of US healthcare. It is that the United States has a strong emphasis on patient-doctor relationships. In this aspect, the doctor can recommend a product regardless of its price. CMS prohibits the doctor from considering costs in most coverage decisions as CMS wants to preserve the patient-doctor relationship, while ensuring that patients have reasonable access to new, innovative technologies (7).

Economy of Medical Device. Other than the cost of medical devices, the economy on the use of medical device can be in the form of saving time, changing outcomes, providing safety and increasing utility. These economic factors play an important role in the value and cost of healthcare. The impact of these factors on the pricing of medical device is examined in greater detail by Kurt H. Kruger, a Wall Street analyst, with coronary stents as an example (4). First, as a patient progresses through the diagnosis, work-up and treatment (including the use of the cardiac catheterization lab, where interventional cardiology procedures are conducted at a cost per minute that can be as high as $500). Thus it is understandable that there would be significant demand for a medical device that would save time in such a setting or improve the outcome of such procedures. Consequently, this situation allows medical device makers to arrive at prices that are associated with the savings they generate or the outcomes they produce rather than on cost basis, leading to high gross margins and above-average profits as illustrated next.

The pricing set for coronary stents (a bare-metal wire mesh to prop open blood vessel) was about $1,000 each and the cost of implantation and other cares would be roughly $10,000. By 2002, over 1.2 million stents were implanted in the US. This sales mean that the 2002 market for stens is about $1.2 billion and the healthcare cost on stent implantation about $12 billion. Soon after, it was found that vessels containing these bare metal stents often reclog, resulting in costly return visits to the cath lab for follow-on procedures. It was proposed that coating these wire mesh stent tubes with heparin, an anti-inflammatory and anti-coagulated drug, could reduce or eliminate the reclogging problem. Animal and human studies showed dramatic results and the first product, one from Johnson & Johnson, received FDA approval in early 2003. How would the new, improved stents be priced?

Kruger described the pricing calculus this way. First, if a repeat cath lab visit is eliminated, then the saving would roughly be $10,000. If the chance at which the reclogs occurred was about one in three, then the saving from the elimination of one cath lab procedure is $3,300. Johnson & Johnson launched its "Cypher" drug-eluting stent in March 2003 with a price tag of $3,200. Even it is difficult for an outsider, peering into company's financial statements, to ascertain the magnitude of the cost of coating, Kruger, with some insight into the production methods involved, suggested that the coating cost is roughly $75 to $175 per stent to J&J. This cost estimate suggests that the stent companies that provide this improvement would be benefited handsomely. Overall, the innovation first causes an abrupt boost to what is already a very attractive gross and operating margin for the use of stents. Second, the pricing of medical device needs not to be made on cost based. These two observations lead Kruger to conclude that this is the magic of medical technology (and its) economic factors are forceful.

Segmentation. Healthcare is segmented. The needs of young patients may differ from those of senior patients. Thus, your marketing strategy needs to be segmented to suit the need of the various populations. Each state may have a different healthcare policy and offer different incentives to users and payers of your product. To optimize your marketing operation, you need to regionalize the allocation of your marketing resources. Similar consideration should be made when you expand your operation internationally.

Franchise, Consultative Selling, Feedback and Marketing Expenses. According to Kruger, a medical device is known by its sales reps. A company can gain a market share by hiring an established and well regard sales rep in a territory. These reps are highly skilled and trained medical expert who can even "scrub-in" and accompany the physician in medical or surgical procedure. A sale rep whom is paid $250,000 to

$650,000 per year may generate well over $1 million in additional revenues for the company. Kruger further asserted that the sales dollars carry 30% plus operating profits, the expenditures incurred in acquiring sales personnel and their attendant market share are well worth it.

To make the sales of your new devices and procedures to the physicians, the rep must be able to explain their use. With pacemakers as examples, Kruger said first that these devices are implanted to the patient's chest with dozens of "blank" features. Held over the chest of the patient's chest, a wand (connected to a laptop computer) can then program the pacemaker to meet the particular needs of the patient. This "daunting" task is usually undertaken by the physician with the sales rep close at hand serving as a guide. This task is made more acute as medical devices are frequently overhauled or upgraded. In essence, the rep has an additional task on working with the physician on the "how" of using your device while he/she is one of the decision maker on the purchase of your device.

Table 17.1. A Comparison of Marketing and Sales Practices, Historical versus 21st Century.

Historical	***21st Century***
Physicians and Hospitals	Multiple Customers
Physician-Focused Teams	Teams with Focus on Patients, Payers, Prescribers and Policymakers
Hard Sale	Well-differentiated Products and Evidence-based Sales
One Size Fits All	Segmentation Based on Functionalities and Geographies
Sales Representatives	Specialists
Prescribers-focused Messaging with Emphasis on Efficacy and Safety	Value-based Messaging Targeted to Payers, Policymakers and Key Opinion Leaders
Low Strategic Communication Content	High Strategic Communication Content
Advertisements	Advertisements and Digital Marketing
National	Segmentation, Regional, National and International
Cost and Benefit Pricing	Evidence-Based Pricing, Service and Quality
Silo Approach with Some Collaboration between Marketing and Sales	Collaboration among Management, Engineering, Marketing, Sales and Servicing

Another important aspect of the sale rep is the feedback of information from the clinics back to your company. The sales rep, as the company's "eyes and ears," is continually observing the activities of their physician buyers/customers. Because of the rep's financial interest in the company, he/she will actively solicits ideals from the physician and communicate them to your development group to translate those ideas into new products that he/she can sell.

Finally, Kruger also analyzed the marketing expenses of Medtronic, Boston Scientific, Guidant and St. Judes and found that their selling general and administrative expenses come to 19% to 25% of their revenue and their marketing expenses 6% to 8%.

Key Market Characteristics. To highlight the points made in this section, we summarize the key characteristics of marketing practices in Table 17.1 by contrasting what companies did in the past with what they do in the 21st Century.

17.3. Customer Relationship Management (CRM)

CRM is the management of customer information and its flow for your marketing and sales activities involving current and prospective customers (1). As illustrated in Fig. 17.3, the personnel in management, marketing, sales, servicing and engineering utilize the information from

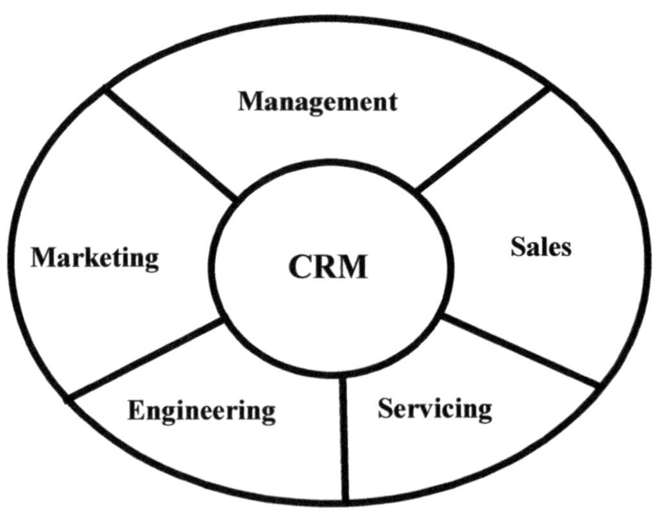

Fig. 17.3. People using CRM and collaboration for success.

CRM to work collaboratively to assure that all customers are satisfied with your company's products and services (1). The CRM software processes and organizes information about customers and customer interactions for easy access by various company departments. The goal of CRM is to enable the company to gain an understanding of the needs of customers in order to obtain new customers, retain current customers, conduct targeted marketing and generate new products.

Communication is the key to CRM. Every interaction with a customer should be treated as an opportunity to make a new sale, build loyalty and emphasize the benefits of your products. To best manage customer relations, it is imperative that you integrate all your departmental efforts from the perspective of customers. An analysis of customer interaction by CRM can generate business insights, drive customer demand and assure a good ROI.

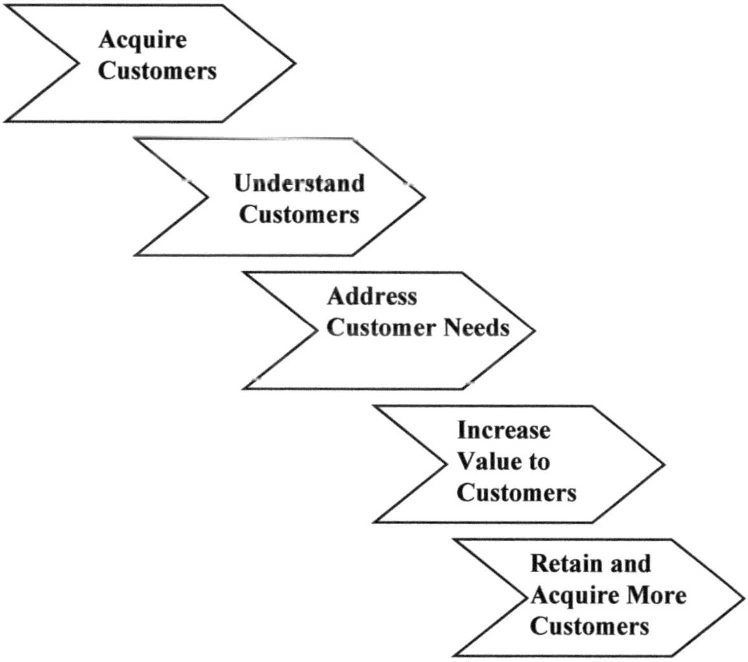

Fig. 17.4. Life cycle of customer relationships

The life cycle of a customer relationship involves the five steps depicted in Fig. 17.4. You acquire new customers or patients for your product when it is introduced. You need to promote the benefits of your product to patients and illustrate its cost and effectiveness to customers. Similarly, you will examine whether your product really address the needs of customers. In the meantime, you will work to increase the value of your product to customers. By showing the benefits of using your product and its high use value for patient healthcare, you will have a good chance to retain your current customers and acquire additional new customers.

A CRM initiative is successful only because it:

- Focuses on one specific strategic business issue
- Has clearly defined responsibilities and measurable goals
- Has appropriate organizational structures and team collaboration
- Sets clear customer strategies

Your CRM is bound to fail if you want it to revamp your entire business or to cover the entire spectrum of the life cycle of customer relationships.

To assure the success of CRM, you need to be diligent with the following assignments to company personnel:

- Management: Assurance of collaboration and goal setting on CRM
- Engineering: Design of cost/effective products and the development of new products
- Servicing: Responsive service for customer retention and intelligence gathering in the field
- Marketing
    - Messaging, workshops, education programs and digital marketing
    - Segmentation and targeting
    - Advertising
    - Promotional materials
    - Interaction with Key Opinion Leaders
    - Gathering patient stories
    - Public relation
    - Competitive intelligence

- Sales
  - Training programs for sale representatives to become specialists
  - Relationship selling to physicians, hospitals and managed care organizations
  - Marketing to professionals

17.4. Marketing Ethics and Legal Compliance

Marketing ethically is now a key concern of all medical companies. Because of the large retainer payments or gifts, which they gave to surgeons, five major orthopedics companies were fined millions of dollars for the infringement of Federal anti-trust laws (3). These companies have been ordered to "clean up their act." Legal observers have been assigned to each company to scrutinize their marketing activities to ensure strict compliance with the law. The importance of legal compliance was further highlighted by a headline news item in September 2009 that reported that the Feds had fined Pfizer $2.3 billion and called the drug maker a repeat offender of the illegal promotions of 13 drugs, including Viagra, Zoloft and Lipitor.

Several states have enacted laws to regulate the marketing activities of medical device and pharmaceutical product companies. Codes and guidance on the conduct of these companies have been released by a number of trade associations and the Office of the Inspector General, DHHS. The laws and codes encourage or require the establishment of a Comprehensive Compliance Program to address concerns about the interactions between companies and healthcare professionals and to address the public and private sectors' desire to reduce fraud and abuse (2). A strict limit has been set on gifts, promotional materials items or activities that a company might give to individual healthcare professionals.

There are some exceptions to this rule. One is that drugs samples can be given for free distribution to patients. Another is that financial support for education forums can be given only for those that are operated by an independent entity. In addition, the compliance program of your marketing and sales plan must be transparent to the public.

17.5. Preparing Your Marketing and Sales Campaign

By this time, you should have some understanding of the complexity of the market for medical devices and pharmaceutical products. While preparing your marketing plan, you should also evaluate whether the marketing of your product is best done by an in-house marketing and sales team or through a well-established distributor. If you elect to do it within, there are three things to consider doing as you move your young company forward with your own marketing and sales efforts:

- Have a patient satisfaction survey done and collect evidence on the cost, effectiveness, benefits and safety of your product. Use health models to project the economic benefits from the use of your product. The survey, evidence and models are organized for review and evaluation by prescribers, payers and policy makers.
- Recruit a VP of Sales and Marketing who is well-qualified, experienced, and knowledgeable of the healthcare that you intend your product to serve. If you intend to sell by a distributor, make sure that its sale manager is qualified as if he/she is, in effect, your VP of Sales and Marketing.
- Hire qualified sale representatives and set goals and incentives for them to work with. Train and equip your representatives with the knowledge and presentation required of a specialist.

Once you have built up a customer base, consider the following list for implementation in your expanding marketing and sales campaign:

- Acquire CRM software for information processing, set up the process to collect customer information, and provide resources and incentives to assure collaboration among management, engineering, marketing, sales and services.
- Make information on your product available to key opinion leaders so that they can inform the medical community of the cost/effectiveness of your product.
- Set up your website with information on causes of diseases, treatment options, your product's technology, and stories of patients.
- Establish education programs, web training and web courses for physicians so they may learn about your product and get in touch with them through relational contact.

If you elect to use a distributor to market and sell your product, go over the prospect list with the distributor and discuss with him/her how prospects on the list should be contacted and who will do it.

Once you reach the mature stage and begin to achieve a satisfactory level of sales, you need to commit to the long term and continual upgrading your products using input from customers and bringing your marketing and sales campaign to a new high. The following commitments and actions are suggested for your consideration:

- On commitment and customer knowledge:
  - Allocate resources in gathering hospital intelligence on purchasing and product needs
  - Align hospital purchases with product development, marketing, sales and servicing
  - Implement a lifecycle approach in the commercialization process
  - Build collaboration among production, marketing and sales personnel to integrate the knowledge of hospital needs, trends, processes, and economics into the development, marketing and sales effort.
- On action development:
  - Effectively integrate hospital needs into your company's R&D
  - Gear the evidence generation processes toward all customers with special consideration to hospitals
  - Incorporate the product portfolio and positioning strategies with hospital economics and adaptation risks
  - Incorporate hospital needs and product requirements into marketing materials and the training of sale representatives
  - Develop and deliver education materials and servicing support to hospital customers
  - Match product positioning, pricing, and contracting strategies with hospital economics and adoption risk
  - Put into place an account planning process to evaluate and develop a marketing plan for hospital consumers and physicians
  - Train sales representatives and management staff in the tools, tactics, knowledge and ethics applicable to hospital's material management and purchasing processes

- Have sales representatives actively engage hospital customers in product launches and on-going sales activities.

17.6. Concluding Remarks

The spiraling increase in healthcare expenditures in the 21st Century and the availability of so many medical devices and pharmaceutical products for diagnosis and treatment have put pressure on everyone engaged in sales of these products. You can become the leader in the market for your specialty product if you have a quality and cost-effective product that stands out from the competition and your marketing campaign can communicate the benefits of your product to customers.

References

1. Arnold, J., *Future Pharmaceuticals*, Q3, (2006).
2. Buckley M. D., and B. J. Flynn, New California Law Required Comprehensive Compliance Programs for Pharmaceutical and Medical Device Companies, www.nutter.com
3. Kidwell H., www.medicaldevice-network.com/features/feature40463, Aug. 2008.
4. Kruger, K. The Medical Device Sector, in *The Business of Healthcare Innovation*, ed. L. R. Burns, Cambridge Univ. Press, pp. 271-321 (2005).
5. Medtronic 2007 Annual Report.
6. National Library of Medicine, Health Technology Assessment 101, www.nlm.nih.gov/nichsr/hta101/ta10104.html, March (2008).
7. Novelli, T., *Med. Device Tech.*, May/June 2009.
8. Peckham, B., *KidneyTimes.com*, June 2007.
9. Ressi, M. A., www.manhattanresearch.com/presentationresearch
10. Rubin, R., *USA Today,* http://www.usatoday.com/news/health/2009-08-23-dialysis_N.htm?POE=click-refer (2009).
11. Wikipedia, en.wiklpedia.org/wiki/Pay_for_performance_(healthcare)

Chapter 18

Expanding and Globalizing Your Business

Globalization during the last thirty years, the result of freer travel and trade, has changed the world's economy in many ways. The first section of this chapter is a description on the disease prevalence of the world. We then examine the healthcare systems of the UK, Germany, France, Italy, and China. China's effort to build up its R&D and biomedical industry is reviewed. The challenges for you to market your products internationally are discussed. Finally we present some actions you can take to enhance the competitiveness of your product, generate new product and expand your business.

18.1. World Prevalence of Diseases

The prevalence of end-stage renal disease (ESRD), chronic obstructive pulmonary disease (COPD) and infectious diseases throughout the world is examined. In Table 18.1, we show for four countries, USA, Japan, Taiwan and China, the number of ESRD patients taking dialysis and the annual cost of hemodialysis treatments per patient. For comparison, the country's GDP per capita appears in the table's fifth column.

Table 18.1. The prevalence of hemodialysis and its cost of treatment.

Country	Prevalence* (per million)	Dialysis Patients	Annual Cost for HD treatments	GDP per capita
USA	1,300	400,000	$72,000	$46,716
Japan	2,000	280,000	$50,000	$38,559
Taiwan	3,040	70,000	$20,000	$17,040
China	2,000	2,600,000	$6,400	$3,315

* The prevalence in Japan is used to estimate the number of patients in China who require hemodialysis treatments.

Disproportional expenditures like that of ESRD are bound to show up for the treatment of chronic obstructive pulmonary disease (COPD) because too many Asians are heavy smokers. For example, there are more than 320 million smokers in China (67% of its adult males and 4% of its adult females). In 2000, the total cost of COPD and other smoking-related diseases in China amounted to more than $5 billion and COPD was the second leading cause of mortality in China, trailing only cerebrovascular disease.

These COPD patients are bound to become short of breath and require assistance in breathing. The sheer number of COPD patients indicates that the impact of COPD on healthcare may be even more severe than the SARS epidemic. During the SARS epidemic, many hospitals ran out of ventilators to provide ventilatory support to SARS patients. Although it may be absurd to assume that each smoker will eventually acquire a ventilator, the price tag for 320 million ventilators at today's lowest price of $4,000 per ventilator comes to $128 billion or 3% of China's 2008 GDP.

"Infectious diseases are the second leading cause of death and the leading cause of disability-adjusted life years worldwide and the third leading cause of death in the United States. Among these infectious diseases causing death worldwide, acute lower respiratory tract infections, HIV/AIDS, diarrheal diseases, tuberculosis, and malaria predominate" Anthony Fauci, Director of National Institute of Allergy and Infectious Diseases said (2). "The sequencing of human and microbial genomes and advances in functional genomics will underpin significant progress in many areas, including understanding human predisposition and susceptibility to disease, microbial pathogenesis, and the development of new diagnostics, vaccines, and therapies."

The healthcare problems of patients with diabetes, obesity and cancer are daunting. We need to extend the "**One World One Dream**" vision of the 2008 Olympics to citizens of the world by encouraging manufacturers of medical devices and pharmaceutical products and governments to work together to achieve the dream of resolving the many health problems that endanger mankind.

18.2. Healthcare in the UK, Germany, France and Italy

Each country in the European Union operates its own national health system. For example, healthcare in the UK, which has a population of

61 million, is provided primarily by the National Health Services (NHS) with an annual budget of about $160 billion (1). Almost all treatments and drugs that NHS offers are free of charge. England does have private insurance that is paid by employers and individuals to supplement their healthcare at a cost of around 10% to 20% of total healthcare expenditures. 92% of in-patients have stated that they are satisfied with their treatment and 87% have said that they are satisfied with their general practioners.

In 2004, the NHS began a major pay-for-performance initiative that is known as Quality and Outcomes Framework. General practitioners receive a 20% funding increase from the previous level to invest in staff and technology in order to increase the use of electronic health records and the electronic prescription service. The data has shown that the pay-for-performance has improved the healthcare quality for 10 chronic diseases.

Germany has an insurance-based healthcare system for all of its residents. 90% of residents are enrolled in public, non-profit health insurance plans. They have a free choice of health insurance and health providers: preferred provider organization (PPO), health maintenance organization (HMO) and primary care. There is one standard list of benefits for all insured persons.

The healthcare systems of France and Italy were ranked 1st and 2nd by World Health Organization (WHO) in 2001 because of their universal coverage, responsive healthcare providers, patients and provider freedoms and the health and longevity of the population. The French law protects a patient's choice of practitioner and guarantees the physician's control of medical decision-making. The French government grants the private insurers the right to run its healthcare funds.

The French system makes it more difficult for insurers to deny coverage for preexisting conditions or to those who are not in good health. At $3,500 per capita, the French healthcare system is one of the costliest in Europe, but still far below the cost of $6,100 per person in the US.

The Italian healthcare system is a decentralized version of the UK's NHS. Despite the WHO's high ranking of the Italian system, only about 20% of Italians are satisfied with the quality of their care.

18.3. The Healthcare Systems and Biomedical Industry of China

Characteristics of China's Healthcare. When the Chinese Communist Party took power in 1949. The government owned and operated all health care facilities and employed the health care workforce. No health insurance was necessary, because services were nearly free. In 1984, China turned its economic system to a market economy and reduced the role of government in all economic and social sectors, including health care. By 2008, China's leaders had concluded that major reforms in both insurance and the delivery system were necessary to shore up the system and make a commitment to providing affordable basic health care for all Chinese people by 2020.

Some economic and health profiles of China and the USA are given in Table 18.2. First the health expenditure per capita of the USA is much greater than that of China. However the similarity in life expectancy and under-5 mortality rate indicate that the quality of health care in China does not differ greatly from that of the USA. In terms of chronic diseases, China does have problems because of its large smoker population, whereas the USA has a large obese population.

Table 18.2. Economic and health profile of China and USA*

	China	USA
GDP per capita, 2016	$8,123	$57,467
Population using improved water	90%	99%
Health expenditures per capita, 2014	$410	$9,300
Life expectancy	75 years	79 years
Under-5 mortality rate, 2015	1.1%	0.7%
Population over age 60	14%	20%
Smokers	24%**	15%
High glucose	9.5%	10.3%
High blood pressure	29.2%	15.6%
Obesity	5.5%	31.7%

* Data are for 2008, except where noted differently. Data are from WHO.
** Fraction for male is 47% and that for female 2%.

The healthcare in China is provided by the village clinics and three tiers of hospitals:

- ***Village Clinics.*** The government encourages hospitals to own and manage the clinics that are rated more satisfactory over those that are owned and managed privately or by village committee.

- ***Primary Hospital*** (Tier 1). Typically, this is a small hospital that contains less than 100 beds and serves a small town. It provides preventive care, basic healthcare and rehabilitation services.
- ***Secondary Hospital*** (Tier 2). This serves a medium-size city, county or district and contains up to 500 beds. It provides comprehensive health services. It also offers medical education and conducts research on a regional basis.
- ***Tertiary Hospital*** (Tier 3). This is a general hospital, typically in a large city, and has more than 500 beds. It provides specialist health services, plays a leading role in medical education and scientific research and serves as a medical hub that provides care to multiple regions. Most major pieces of equipment in these hospitals were imported from the USA. The operational characteristics of the Tier 3 hospital are shown in Table 18.3. It is noted that the total project cost is much lower for the Chinese hospital, but its normalization by GDP per capita is higher. This implies a greater willingness by the Chinese government to build up the healthcare infrastructure. However, the length of stay in Chinese hospital is much longer and their staffing is lower.

Table 18.3. Operational Characteristics of a 500-Bed Chinese Hospital and International Hospital in 2008 (8)

Chinese Hospital	Factor	International Hospital
500	Beds	500
46,500	Square Meters Occupied	93,000
$23,250,000	Cost to Build	$218,000,000
$70,000,000	Total Project Costs (TPC)	$350,000,000
700	Total Employees	2,000
300	Physicians	500
200	Nurses	600
200	Other Staff	900
1.4	Employees/Bed	4.0
17.5 days	Length of Stay	4.5 days
3:1	Clinic Visits/Inpatient Day	17:1
63%	Occupancy Rate	68%
18,396	TPC/GDP per capita	7,231

In 2008, the government defined its approach to electronic health (e-health). Since then, hundreds of e-health record systems have been developed, although the systems are still in their infancy.

The income of urban residents is five times that of rural residents. Urban health insurance was initiated in 1999 to provide broad coverage for urban residents. The insurance cost is shared by employers and individuals, with each worker paying 2% of his or her wages and employers paying 6%. In rural areas, although farmers pay a small fee and public funds pay the balance, only 25% of China's rural population is covered by rural health insurance. As a result, all rural primary health services are provided on a fee-for-service basis.

Overall, China is a developing country with healthcare services for the urban and rural population (52% vs 48%) that are not equal. By 2012, the government had subsidized a modest, but comprehensive, insurance system for 95% of the population.

Pharmaceutical and Biotech Industry. Spending on pharmaceuticals in China exceeded $107 billion in 2015, versus $26 billion in 2007. The country is projected to become the second-largest healthcare market by 2020.

China now has many R&D centers and a critical mass of leading scientists and interested investors. As exemplified by the developments described in this section, China is poised to achieve many more pharmaceutical innovations in the near future.

Xian-Ping Lu left his job as director of research at Galderma R&D, a drug manufacturer in Princeton, NJ and founded with a group of US-trained Chinese scientists Chipscreen Biosciences Ltd. in Shenzhen in 2001. In February 2015, Chipscreen announced that its first drug, a medication for a rare type of lymph-node cancer, had been introduced to the market in China. This drug, which is called chidamide, or Epidaza, was developed from start to finish in China. It is the first oral drug to be approved for sale in China for the treatment of peripheral T-cell lymphoma, and only the fourth in a new class globally (3). Lu estimates the research cost of chidamide to have been about $70 million, or about one-tenth of what it would have cost to develop in the US. Chipscreen raised $43 million in five rounds of venture funding and more than $20 million from the Chinese government through programs, such as the Thousand Talents Program. The latter program brings overseas Chinese scientists and engineers back to their home country to launch their own companies and to work in scientific organizations and universities.

Chidamide is now on the market in China for $4,275 a month, a price far lower than patients in the US pay for some of the newest cancer medicines, but much more than the typical Chinese patient pays for drugs. Lu said that the price reflects a balance between affordability for patients and a return for shareholders, some of whom wanted to price the drug higher. The website of Chipscreen identifies an Aipusha charity project from which qualified patients can obtain Epidaza without charge.

China is building up many R&D centers, creating a critical mass of leading scientists, promoting entrepreneurship and innovation and generating venture investment and government support. The biomedical industry is taking advantage of these supports to expand its R&D.

One example is the biotech, BeiGene, which gained ground in the drug-discovery process by deploying a large-scale drug-testing team to testing compounds on human tissue (such as cancerous tumor samples) during the preclinical phase to obtain early indications of issues that might arise in human testing, and capitalizing on access to China's large pool of patients. In July 2017, investor/trader Jonathan Faison concluded the following (5):

- Celgene collaboration terms (with BeiGene) were quite generous and could be indicative of further upside,
- BGB-3111 has the potential to be a best-in-class BTK inhibitor,
- BeiGene is the only company to test a wholly-owned PD-1 and BTK inhibitor combination

In genomics research, a state owned enterprise, Beijing Genetic Institute (now known as BGI-Shenzhen), is deploying massive resources (2,000 PhDs and more than 200 gene-sequencing machines) to solve biotech problems. Christina Larson of *MIT Technology Review* referred to BGI as China's gene factory (7). In 2010, with the aid of a $1.58 billion line of credit from the China Development Bank, BGI purchased 128 state-of-the-art DNA sequencing machines for about $500,000 each. It now owns 156 sequencers from several manufacturers and accounts for some 10 to 20% of all DNA data produced globally. So far, it claims to have completely sequenced some 50,000 human genomes—far more than sequenced by any other group.

BGI's sheer size has put Chinese gene research onto the map. In 2002, BGI published the complete sequence of the rice plant in *Science*. Research in the national interest of China has been a BGI mainstay. It decoded the DNA of the giant panda, and also the genetic mutation that makes Tibetans so well suited to life at high altitudes.

The scale of BGI could be an advantage as comprehensive gene readouts become part of everyday medicine. The cost of DNA sequencing is falling quickly. In a few years, it's likely that millions of people will want to know what their genes predict about their health. BGI may be one of the gene factories to tell them.

In conjunction with the growth of pharmaceutical industries, China also has become a mecca of clinical trials. The prospects of medical product commercialization in China are massive and many international pharmaceutical companies build up their research centers in China to conduct R&D and clinical trials. Clinical trials are the scientific soul of cancer-cure advances. However, nearly 20% of these trials in the USA fail to enlist sufficient patients to render verdicts on experimental therapies. Because of the ease of patient recruitment in China, many trials are now conducted there. Equally important is that the standards of clinical trials in China have moved closer to international standards. This encourages the development of innovative drugs and conducting their clinical trials in China. According to www.clinicaltrial.gov, a website that the US government operates, 5,552 studies have been registered throughout the world, of which 5,045 involved China (i.e. 90%).

Growth of Medtech in China. In 2016, the medical device market in China reached a value of $54 billion, an increase of 20% from that in 2015. 73% of this growth is fueled by hospital procurement in China. In comparison with other sectors of the Chinese economy, the market in medical devices is one of the fastest growing market sectors in that country.

China relies heavily on foreign imports for its medical supplies and devices. This is especially true for high-tech, high-priced items, of which the US, Germany and Japan are major exporters. For example, 80% of CT scanners, 90% of ultrasound imaging equipment and 90% of MRI equipment are manufactured abroad. All of these devices are bought by tier 3 hospitals (those in metropoles) and 66% by tier 2 hospitals (those in larger cities). Imports of mid-to-high end products by China have declined in recent years, due partly to foreign companies moving their production plants to China.

This characteristic difference between foreign and Chinese producers of medical devices is attributed to two reasons. First, there are high entry barriers for the higher-end medical device market as their development is capital-intensive, has a highly technical basis and is lengthy. Secondly, the Chinese prefer to use well tested, quality products that are expected to have the least chance for malfunction. In addition, imported medical devices tend to sell in China at prices 50-100% higher than in the countries where they are produced.

China has several thousand local manufacturers of medical devices. They are mostly small companies with low revenues, 90% of whom are makers of low-tech products, such as syringes, bandages, surgical gloves, thermometers, blood pressure monitors and pulse oximeters. Both the export value of these low-tech products and the import value of high-tech products are comparable and ranked fifth among various industries.

Since 2009 several companies with medium technology have merged and gone on to take larger shares of the market. These include the Wei Gao Group, Mindray, Yuwell, Microport, Le Pu Medical and Fosun Pharma. Thousands of employees were laid off in these mergers. These companies are quite profitable now as they can sell their products at lower prices than their foreign competitors.

Electronic Health Records (EHRs) are quickly becoming commonplace in China. The three internet giants of China - Alibaba, Tencent and Baidu - are actively involved in the improved operation of EHR and mobile health. International technology companies, such as Apple, Google, Dell, Siemens, IBM, Fujitsu, and Cisco could expand their presence in EHR and mobile Health.

The aging of the population and rising living standards in China have given rise to a rapid growth in the demand for medical products and services. Despite a 20% increase in public spending on health care in recent years, China currently makes up only 2 to 3% of the global market for medical devices. Internationally, the respective markets for medicines and medical devices are approximately equal in size. However, the market for medical devices in China is only 14% of that for medicines.

China's R&D and International Initiatives. China has risen rapidly to become the world's second-largest performer in R&D as it accounts for 20% of 2016 global R&D expenditure, according to the US National

Science Board. The US still leads R&D with a 27% share, which is down from almost 33% a few years ago.

The US is strongest in biomedical research, clinical trials and the number of articles published in high and mid-ranking journals. This position is being threatened as China reforms its science and technology system and provides major career incentives to authors for the publication of articles in top-tier western journals.

China is now invested heavily in building up a network of science and technology parks and offers support to businesses that need to convert the results of research to products and services. There are now 17 national high-tech innovation parks and 146 regional ones across the country. By early 2016, China had already more than 5,000 high-tech start-up incubators. For example, manufacturing centers in Suzhou are on their way to becoming an innovation hub. They use their own resources to attract topnotch high-tech companies, such as Apple, to build a new R&D center in Suzhou.

In 2015, Chinese President Xi Jinping pledged to create 100 health projects that was designed for women and children in the developing world. The country has also finalized a contribution to global health through its Belt and Road Initiative—a massive cross-border transportation project linking China to Eurasia—by forging a deal with the World Health Organization.

"(China) has a history of more than 50 years of cooperation with, and aid to, Africa. One of the most recent instances of this support was its dispatch of Emergency Medical Teams to West Africa to assist in the Ebola outbreak. In addition to its support in emergencies, China's capacity to manufacture high-quality, safe, and effective vaccines, medicines and medical devices promises to become a critical global resource for health," The Director General of the WHO, Tedros Gebreyesus, said.

The US CDC and the Chinese CDC are working to build an African CDC headquarters in Ethiopia and five additional centers throughout the continent to help to fight infectious diseases. China is providing $2 million and the United States is contributing $10 million to the venture.

In May 2017, Bill Gates traveled to Beijing to open the Global Health Drug Discovery Institute — a partnership between the Gates Foundation, the Beijing municipal government and Tsinghua University — that will focus on early drug discovery to combat diseases, such as tuberculosis, malaria and HIV. The three entities are providing more than $100 million to support its operation in the first five years of operation.

18.4. Global Markets of Medical Devices

Overview. As a percentage of GDP, the healthcare expenditures of the Big Four of the European Union are about half of US expenditures. The World Health Organization (WHO) ranked these countries high on the quality of healthcare. China is already the world's manufacturing center, can the US maintain its leadership position in the medical device industry? The BRIC (Brazil, Russia, India and China) are the largest and fastest growing countries in the third world. With more purchasing power than the USA for this 42% of world population, BRIC is poised to become a large market for medical devices.

US Export to the World. The total value of US medical device exports in 2007 was $32 billion with almost 46% of the devices going to the European Union. About 22% went to Japan and other Asian countries. The industry for medical devices is one of the few USA industries that export more than they import. In 2015, the US exports increased to $43 billion.

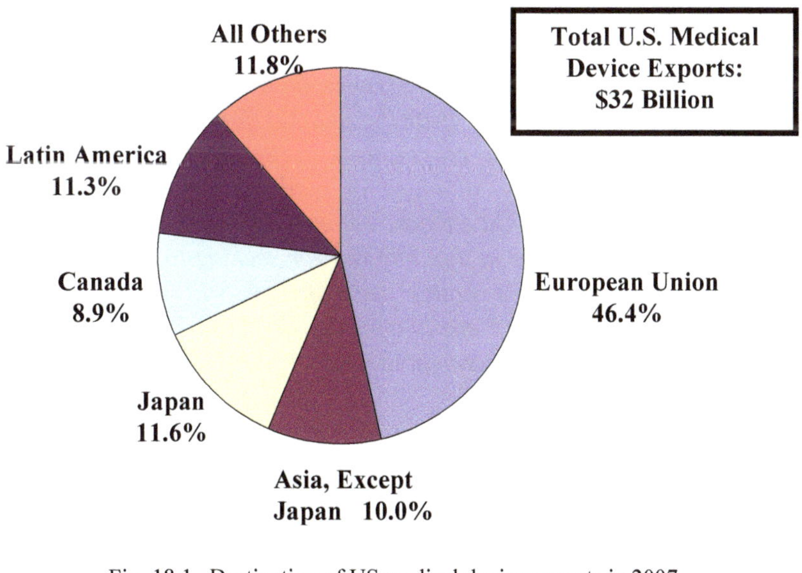

Fig. 18.1. Destination of US medical device exports in 2007.

Global Market for Medical Devices. According to the International Trade Administration (ITA) of the US, the global market for medical devices had a value of $340 billion in 2016. It was $210 billion in 2008. The medical device market shares of five regions of the world in 2016 and the 2020 forecasts are listed in Table 18.4. The US share of the world market has remained at roughly 45~49% in recent years.

Table 18.4. Medical Devices Market: Forecast for Growth, in $billion

Region	**2016**	**2020**
Americas	166.6	208.6
Asia/Pacific	68.7	88.6
Central/Eastern	14.6	19.1
Middle East/Africa	10.0	13.2
Western Europe	79.5	106.2
Total	**339.5**	**435.8**

The Chinese market for medical devices was forecasted to reach a value of $23 billion in 2011. Currently, it is the second largest medical device market in Asia. China has the potential to become the largest single-country import market for US medical device exporters. The production and development of high-tech medical equipment in China continues to be largely small-scale. Fierce competition describes the market in China for medical devices and supplies.

In April, 2008, C. A. Padilla, Under Secretary of Commerce for International Trade, led 15 US companies on a healthcare trade mission to Beijing. "*China is a fast-growing market for the U.S. healthcare industry, known for its world-class goods and services. U.S. industry has the potential to raise the standard of medical care in China,*" said Padilla. "*China's pending healthcare reforms will affect many American companies doing business in that market—or seeking to do so. A clearer understanding of China's policy environment will enable the U.S. healthcare industry to more easily compete in China, raising the quality and efficiency of medical treatment there.*"

18.5. Challenges to Global Marketing

Government Support. In global trade, ITA) is the key US government agency that is engaged in strengthening the competitiveness of US industry, promoting trade and investment and ensuring fair trade and compliance. ITA's Office of Health and Consumers Goods has a Health Products and Technologies Team to provide information about foreign markets to industry. As marketing opportunities in medical device develop, ITA targets four areas — quality, regulatory structure, pricing and reimbursement and IP issues with special emphasis on China, India and Brazil. The movement of manufacturing to overseas locations can lead to the introduction of low quality products, which will weaken public confidence and affect trade adversely. The ITA is working with the FDA on industry training, adoption of good manufacturing practices and counterfeit detection.

ITA encourages international transparency, fairness and harmonization to ensure that there is fair trade in world markets for medical devices. ITA also discourages the establishment of trade barriers by discrimination against imports, the use of foreign reference pricing and an excuse of cost containment. The ITA) can assist you to overcome barriers related to the following:

- Excessive tariff and custom regulations (reporting it to www.export.gov/tcc, TCC: Trade Competitive Center)
- Discriminatory rules of origin
- Burdensome standards
- Lack of IP right protection (filing a complaint with www.stopfakes.gov)
- Non-transparent government procurement.

As discussed previously, the SBA is very supportive of globalization. Its affiliates conduct courses on global enterprises, exporting and importing to help your staff learn more about globalization.

Many European governments are taking a more active role to support entrepreneurial developments. For example, if you decide to develop a high-tech company in France and are looking for VC funding, the French government will guarantee 75% of the VC investment. The governments of many Asian countries are investing in the development of their medical device and biotech industry. Government investments made by China are detailed in Section 18.3.

Foreign Patents and Counterfeit Prevention. The legal protection afforded to a patent that has been approved by the USPTO applies only to markets in the United States and its territories. The global market position of your medical inventions may justify filing your patent internationally as well.

Most countries are members of The Patent Cooperation Treaty (PCT). A US applicant can now file one "international" application in English to the USPTO. This is not a substitute for patent applications to other countries. What you gain with the PCT filing is an additional 18 months in which to file your patent application to the patent offices of other countries.

For a European patent, you first apply for approval from the European Patent Office. When the patent is approved, you can pay one large lump sum to have all countries in the European Union grant you the patent right. If you prefer, you can pay the patent filing fees for a limited number of large European countries to protect your intellectual property right in those countries. If you want to include the Asian market, you will need to apply for patents in Japan, China, Taiwan, Singapore separately. This multinational process is further complicated by those non-English speaking countries like Japan or China, which only process patents applications that are written in their language.

Counterfeit copies of your products are usually not produced under the good manufacturing practice that is prescribed by the FDA and are of low quality. The counterfeiter may even duplicate your labels with the result that users may send the imitation products to your company for repair or refund. In addition to government effort, you should consider adopting additional protection for your device by making it difficult to copy.

Regulatory Affairs and Reimbursements. To market your device to the European Union, obtaining a classification from the European Medicine Agency is the first step (6). That agency's four classifications, I, IIa, IIb and III, are not much different from those of the FDA. The classifications are governed by 14 rules and 24 articles of the Medical Devices Directive (93/42/EEC). For other products, the Active Implantable Medical Directive (90/385/EED) and In vitro Diagnostic Directive (98/97/EC) may apply.

Once it has been approved by the European Medicine Agency, the medical device agencies of some EU nations review the device classification again and set the fee. You will need to obtain a CE (Comformité Européenne) marking to confirm that the product meets all health and safety requirement of their directives. A "notified body" that is qualified by the EU nations must be involved in the conformity assessment process if you wish to sell your product in the EU.

With a mandate to promote cost-effective technologies and to work on eliminating harmful or ineffective technologies, the Health Technology Assessment Agencies of EU nations work together to facilitate the exchange of information and harmonization. On the other hand, the EU Treaty respects the national right of each member to operate its own healthcare system. EU member nations have independent control of their pricing and payment methods. Although almost of the countries have implemented evidence-based medicine, each country's pricing, market access and enforcement arrangements differ.

To do it correctly, you and your distributor in the EU may have to work with a notified body to collect all appropriate documentation and to ensure that you fulfill all regulatory requirements for submission to the appropriate agencies for classification approval, CE marking (the comformity certification marking of EU) and device reimbursement.

The regulatory agency of Japan is the Pharmaceutical and Medical Safety Bureau of the Ministry of Health and Welfare. For China, you may need to submit your device approval application, if the device is class I or II, to the Pharmaceutical and Medical Safety Bureau of the provincial government. The application for Class III needs to be submitted to the National Pharmaceutical and Medical Safety Bureau. Although China's rules and regulations for medical devices bear considerable similarities to those of the US and EU, data on clinical trials of Class II medical devices, which may not be required by the US FDA, must be included in Chinese submissions.

Complications may arise when you market to countries in Latin America that may not have full-fledged regulatory systems.

Issues on Distribution and Marketing. There are many options available to you to distribute and market your product in a foreign country. You may set up a subsidiary and hire employees to sell the product. Paying commissions to sales representatives or permitting someone to set up a franchise are other ways of generating sales. You could also engage a distributor, set up a joint venture or license a foreign company to sell the product.

In this process, you will sign a distribution and marketing agreement with a foreign distributor to sell and distribute the product. Certain aspects of the agreement are similar to those in the licensing agreement that was examined in Section 14.4. Below are additional points to consider:

- On performance and commitment of the distribution entity
  - Undertake due diligence of the foreign distributor's business plan for distribution. Also, do a credit check, contact all references and review past performance.
  - Determine the extent of the distributor's commitment to the required facilities and staff. What is the marketing obligation regarding efforts and expenditures?
  - What will be the training frequency and who will pay the training expenses?
  - What will be the arrangements about keeping, monitoring and auditing sales records, customer information, and competition?
- On sales exclusivity and performance
  - Consider the exclusivity of a full line of products or a partial line.
  - Negotiate the number of, and investment in, demos, shows and visits.
  - Who does (and approves) the marketing planning?
  - Consider pricing discounts, volumes goals, payment schedules and incentives.
  - What limitations should the distributor be subject to in order to prevent it from offering "competing" products
- On renewal and termination
  - What should be the length of the initial term?
  - Should the contract automatically renew or expire without a confirmation of renewal?
  - Consider the following grounds for termination: breach of contract, cessation of business, bankruptcy, or deficiency in performance.
  - What about notices, resolutions, and a transition to non-renewal or termination?
- On financial arrangements
  - You will need a pricing and payments schedule.
  - Who pays custom duties, value added tax (VAT), sales and other product taxes?
  - Who controls the price schedule, timing, and resale?

    - What about the cost of insurance cost and broker fees?
- On management issues
    - You will need to make arrangements for shipping, warehousing, and the responsibility for product in transit, FOB and FCA.
    - What about the application for import and export licenses?
    - What about necessary product servicing, customer support and warranty management?
    - You will need to limit liabilities and establish caps in both amount and period.
- On ownership of rights and properties
    - Making, manufacturing, copying, modifying and marketing the products
    - Enhancements, documentations, and their translation
    - Records and documentations of property ownership
    - Require timely transfer of properties, documentation and translation materials at termination or non-renewal.
    - Ownership, use and registration of trademarks
- On dispute resolutions
    - What are the governing laws and to which jurisdiction will your efforts apply?
    - What will be the role of mediation and binding or non-binding arbitration?

18.6. Improving Your Products and Expanding Your Business

The competitive environment of the biomedical industry is ever expanding. Even if your company is highly profitable, you need to plan what to do well before your patent expires. Your products can become obsolete when your competitors introduce a safer and more effective healthcare product. In addition to improving the company's operation, the following three action items are worthy of your consideration.

Expanding R&D and Product Lines. The products of your start up are a direct result of your R&D. Further R&D is a long term investment for you to improve your products or to develop new products. How much to invest and how to get the R&D done are too issues that you need to address.

You have three options in getting the R&D done. One is to invest in the team that you already have and the second is to establish a subsidiary to undertake the R&D. Both will represent an investment from your company and require good incentives for your employees or subsidiary to have them transform R&D to products that your customers will want. The third option can be effected by an acquisition, as described later.

In recent years, the R&D productivity challenge has become particularly difficult to overcome in the pharmaceutical sector. Some companies cut R&D expenditures or close their R&D centers to boost their short term earnings. The cost of developing a new drug has increased, whereas the rate of introductions of new medicines that have been approved by the FDA has remained constant at best. In addition, attrition rates have increased sharply, especially in late-phase clinical trials of cancer drugs.

Strategic Alliances. Strategic alliance (SA) is key to the industry in the development of new drug, device or their combination. SA is called for as there is no way that one company can access all the technology it needs if it is restricted within. SA can be in the following four forms (4):

- ***Therapeutic area alliances.*** The partners are allied to focus on a disease category or on disease emanated from a physiological cascade.
- ***Data mining alliances.*** The parties collaborate in the large-scale generation and screening of libraries, for example.
- ***Technology development alliances.*** In one such alliance, one partner receives financial resource from the other partner to complete the technology development and then the other partner brings the technology to the market.
- ***Technology transfer alliances.*** This form of alliance is represented by this working relation: one partner of the alliance obtaining enabling technologies from the other partner for further development and clinical trial.

Merger and Acquisition. You may expand your business by acquisition. For example, Medtronic acquired MiniMed and Medical Research Group (MRG) in 2001. Medtronic manufactures stents, catheters and other devices to treat chronic heart conditions. Acquiring MiniMed and MRG gave Medtronic a stake in the rapidly growing diabetes treatment field and an access to a new "artificial pancreas" that was being jointly developed by these two companies. "We believe adding this new business will provide us a growth vehicle above the corporate average," Medtronic CEO, Art Collins, said.

What is the connection between diabetes and heart disease? According to the American Heart Association, 68% of people with diabetes who are aged 65 and older die from heart disease and 16% die of a stroke. NIH states people with diabetes have additional ways of developing heart disease, are at higher risk of heart disease and may develop heart disease at a younger age.

References

1. en.wikipedia.org/wiki/Healthcare_in_England
2. Fauci, A., Infectious Diseases: Considerations for the 21st Century, *Clinc. Infect. Dis.*, 32:675-685, (2001).
3. Lu, X.P. *et al.*, Development of chidamide for peripheral T-cell lymphoma, the first orphan drug approved in China, *Intract. & Rare Dis. Res.* 5:185–191 (2016).
4. Northrup J, The pharmaceutical sector, in *The Business of Healthcare Innovation* ed. L.R. Burns, Cambridge Univ. Press; pp. 89-100 (2005)
5. seekingalpha.com/article/4088681-beigene-china-based-biotech-continues-surprise
6. www.export.gov/article?id=China-Medical-Devices
7. www.technologyreview.com/s/511051/inside-chinas-genome-factory/
8. www.ucdenver.edu/academics/colleges/PublicHealth/research/centers/globalhealth/news/Documents/Articles%20Gifford/Gifford,%20Chinese%20Healthcare%20Facilities.pdf

Chapter 19

Investing in the Future

The knowledge of community leaders, prominent academicians, authors, government officials, entrepreneurs and philanthropists on the subject of **Biomedical Engineering and Healthcare** that the author has collected from news articles, magazine publications and books in recent years forms the core of this Chapter. The author is especially grateful to Drs. Peter Katona, Michael Khoo, John Linehan and Robert Nerem, who devoted additional effort to writing the excerpts and sharing their vision on investing in the future with the readers. At the end of each section, the author has written a brief conclusion to integrate the information that was presented by those who contributed to that section. Special thanks are due to Dr. Shu Chien who wrote an exciting and passionate Chapter 20 on Biomedical Engineering Entrepreneurism.

19.1. Investing in Your Company and Yourself

On Running a Medical Device Company, Bill Hawkins, Chairman and CEO, Medtronic (Courtesy of *MD&DI*, 5).

1. Core Values: Discipline, Respect, Duty. These are deceivingly simple concepts that seem to recommend themselves to everyone, but are practiced consistently by few. Having been taught these values at an early age, they have played a role throughout my career in medical technology. From my first sales position to leading a great company like Medtronic, I've never found a replacement for core values like perseverance and loyalty.

2. Thinking Differently. Like other high-tech industries, the medical device field requires that we think differently, boldly, creatively, and unconventionally. Challenging norms and assumptions is requisite in order to push the innovation envelope as aggressively as possible. Push

hard—and then some. Don't ever stop. When we get too comfortable, we lose our edge and fail our customers.

3. The Patient, The Patient, The Patient. I strive to stay focused on the patients whom we serve—no matter how challenging or difficult running the business may seem. There's no better way to maintain the passion required for success than to believe in what you can do for your customer. In medical technology, it's simple. It's the patient. The moment I feel the balance shift away from the patients, who are the ultimate beneficiaries of our products and services, is the moment when passion wanes, productivity decreases, and business begins to suffer.

4. Innovation. There's a belief in some business circles that, in challenging economic times, the best companies double-down on marketing in order to win share from competitors. This may be true, but I believe it's even more important to double-down on innovation—the science, research and development behind the new products and technologies that change how healthcare is practiced across the world. Innovation is our lifeblood—and nothing can replace its importance.

5. Technology = Value. Technology for technology's sake is never our goal. A cool gadget may dazzle the eye, but developing truly useful technology to help physicians solve a practical problem provides value and drives market-leading performance.

6. Strive to Discover the Entrepreneur in You. Companies frequently lose their compass and the passion of the entrepreneurial spirit. In our industry, it's critical to stay on the edge of technology. I always think about the future and what medical device advancements might do. What's more, I retain the best engineers and physicians on staff who will share their own visions with me on a regular basis.

7. Family Matters. My greatest sense of pride in my career has been doing what I do each day with my family in mind. The work I do enables many people, including both my father and father-in-law, to enjoy full lives.

8. Have a Mission. Medtronic's mission is nearly 50 years old and has served us well through both good and difficult times. Every company should have a mission to guide its purpose and serve as a reminder of why it exists and who it serves. We all should have individual mission statements as well, to ground and guide us through thick and thin.

9. Live a Little. Working to alleviate the chronic disease burden around the world is serious business. That being said, take some time to live a little and remind yourself of the kind of life you want the patients who use your therapies to have. This reminds you that you're not just saving lives. In many cases, you're enabling better quality, livability, and happiness for those lives.

10. Give a Little . . . Or a Lot. I can think of few better ways to measure one's success than by giving back and giving back often. Those who do will know the powerful and invigorating sense of purpose and satisfaction that comes from helping others and will be humbled by the gratitude expressed in return.

On Customer First, Earl E. Bakken, co-Founder, Medtronic Inc., in *Reflections on Leadership* (1), 2001

For our representatives in the field, of course, "customer first" is more than a strategy or slogan. "Customer first" is the first commandment, the basic law by which they conduct their professional lives. They see or talk with the Medtronic customer every day of the week. They often go to extraordinary lengths to make sure that the customer is satisfied. Our representatives have been known to drive hundreds of miles in the middle of the night to deliver a pacemaker to a physician. They have gone so far as to arrange for the delivery of a competitor's device to a doctor who needed it when one of their own was unavailable. Our field representatives, I am very proud to say, have demonstrated time and again that they will meet the customers' needs regardless of the cost or the effort.

The same might be said of our engineers, our various support staffs, our receptionists, and other company representatives. In the design and manufacture of our products, in our marketing and advertising, in every one of our employees' diverse activities, the prevailing commandment is, or should be, "customer first." Some years ago our employees wore a button that stated the sentiment effusively. The button read, "I Love the Customer!"

On Guiding Vision, R. Jarvik M.D. (the developer of the first permanent total artificial heart) on W. Kolff (6)

In 1971, after my first two years at medical school, Dr. Willem Kolff, who died Feb. 11 at 97, hired me to work on the artificial-heart project at the University of Utah. On my first day, he instructed me to create a new heart design that would keep an animal alive longer than any earlier models had.

Previous designs had failed, he explained, because they did not fit anatomically. And that was all he said. He told me what to do, but not how to do it. That was Dr. Kolff's forte — finding enthusiastic people, laying out his visions and then leaving them to their own devices.

Dr. Kolff, who was one of the founders of the American Society for Artificial Internal Organs, encouraged scores of people to turn their attention to creating mechanical hearts, electronic devices that restore hearing and vision, artificial arms and more. He believed that bioengineering could one day provide a substitute for almost every organ in the body.

What could not be replaced, however, was Dr. Kolff himself, who possessed energy, Old World charm and a grand, guiding vision. I count myself among the many inventors, engineers and doctors who worked with him and will never forget his indomitable spirit.

On Being Successful in Life, Jack Ma (CEO of Alibaba)

Before you turn 20 years old, be a good student.......just to get some experience.

Before you turn 30 years old, follow somebody. Go to a small company. Normally, in a big company, it is good to learn processing; you are part of a big machine. However, when you go to a small company, you learn the passion, you learn the dreams. You learn to do a lot of things at one time. *So before 30 years old, it's not which company you go to, it's which boss you follow.* A good boss teaches you differently.

Between 30 and 40 years old. You need to think clearly whether you want to work for yourself, if you really want to be an entrepreneur.

When you're between 40 and 50 years old, you must do all the things that you are good at. Don't try to jump into a new area. It's too late. You may be successful, but *the rate of dying is too great.* So when you're between the ages of 40 and 50, think about how you can focus on things that you are good at.

When you are 50 to 60 years old, work for the young people. because they can do better than you. So rely on them, invest in them and make sure that they're good.

When you are over 60 years old, spend time on yourself. On the beach - sunshine. It's too late for you to change.

Advice to the young people or entrepreneurs. 25 years old, make enough mistakes. Don't worry! You fall, you stand up, you fall…enjoy it! You're 25 years old. Enjoy the show!"

Thoughts on being a billionaire. I believe that when you have one million dollars, that's your money. When you have twenty million dollars, you start having problems. You worry about inflation and what stock to buy, and so the headaches come. When you have one billion dollars, that's not your money. That's the trust that society gives to you. I believes that you can manage the money, and use the money better than the Government and the others can. Today, I have the resources to do more things with the money and influence that we have. We should spend more time on the young people. I would say, "Someday, I'll go back to teaching." I'll go back to school, spend time with the young people and sharing with them what I've done. The money's not mine. I just carry it happily, and having these resources, I want to do a better job.

On Innovative Company, J. C. Collins and W. C. Lazier, authors of "*Beyond Entrepreneurship*" (3).

To be an enduring, great company, we believe that it must have the ability to innovate continually—to have a constant flow of new ideas, some of which are fully implemented. We say some are implemented (rather than all), because a great company will always have more good ideas than it can fund. Most companies start with a creative founder. The challenge, however, is to become an innovative company, rather than a company that is dependent on an innovative founder. We've identified six basic elements of what it takes to be an innovative company:

1. Receptivity to ideas from everywhere. Highly innovative companies don't necessarily generate more ideas than their less innovative counterparts (good ideas are in plentiful supply for all companies). However, highly innovative companies are often more receptive to ideas—and not only to their own ideas, but to ideas from everywhere.

Furthermore, they do something about the ideas. It is not that an innovative company executes every single idea, but it is much more likely to act quickly on a partly-baked idea than to spend countless hours deliberating about all of the reasons why it can't work.

2. "Being" the customer. One of the best ways to make and keep your company innovative is to have people invent solutions to their own problems or needs. In other words, be your own customer and satisfy yourself. If that is not possible—if you are in a business where you cannot be your own customer—figure out a way to experience the world as a customer experiences it. The idea here is simple. If someone in your company creates an innovation to solve his or her own problem or meet his or her own desire, there are probably other people in the world who would benefit from that invention.

3. Experimentation and mistakes. Sadly, the nature of innovation is that it is fraught with the unknown. The best way to find out if something is a good idea is to experiment, to give it a try. This, of course, leads to mistakes – ideas that aren't good—but they are part of the process. Innovation requires experimentation and mistakes. You can't have one without the other two, period.

4. People being creative. All people have the capacity to be creative. There is no such thing as an inherently uncreative person. Creativity is a capability that resides inside each and every one of us. There is no special breed of person who has been somehow ordained from God with the gift of creativity. Further, it is not true that most of us were born deprived of that blessing.

The first step in having people at all levels be innovative is to believe in their inherent creative capability. After all, how can you possibly expect people to innovate if fundamentally you don't believe that they can?

5. Autonomy and decentralization. This need for freedom and autonomy is well recognized by companies that remain innovative. Herman Miller allows its designers to work away from central facilities in whatever environment they find conducive to work. Merck & Company, one of the most innovative pharmaceutical firms, hires the best scientists that it can find, lets *them* select targets for basic research (not marketing or corporate), and stays out of their way.

This applies to all levels of human endeavor, from a five-person basketball team to entire societies. Indeed, freedom of action-room to experiment—is the primary factor behind the relative strength of western economies compared to their centralized, eastern-bloc counterparts. In their book, *How the West Grew Rich,* Nathan Rosenberg and L. E. Birdzell, show that the underlying source of the West's economic advancement is the large number of innovations that come about by autonomous experimentation. This experimentation, in turn, occurs because people *can* act, because there are few restrictions to prevent them from giving something a try. Just imagine what the United States economy would be like if every new business had to receive formal government approval from a Ministry of Central Business Control. (Actually, we don't need to imagine. Just look at the Soviet economy.)

6. Rewards. We don't mean to suggest that people who are being creative are motivated solely by money, power or prestige. In fact, they're often motivated by the desire for interesting work, the challenge of a difficult problem, the joy of contribution or the satisfaction of doing something new. Nonetheless, innovation should be explicitly rewarded. All people, no matter how pure their motivations, are influenced by the reward systems of their organizations. Rewards matter. So, if you want to remain innovative, you've got to reward innovation.

On Engineering Management, C. M. Chang, Ph.D. Professor of Industrial Engineering, University of Buffalo, State University of New York (2)

Engineers with excellent managerial skills and superior business acumen are needed to lead corporate America in the new century. As the economy grows increasingly global, technologies advance at a faster pace, and the marketplace becomes more dynamic; countless industrial companies will need technically trained engineers to turn technological innovations into profitability.

The need for engineering management training is obvious from another point of view. The National Science Foundation estimated in 2000 that about 46% of American engineers and scientists were actively working in managerial and administrative capacities. This managerial percentage remained more or less constant across the age groups from under 35 to over 55 years old. As the trend continues, almost one out of every two engineers or scientists will be engaged in managing people, projects, teams, technology and other resources to add value to their companies.

Conclusion, Jen-Shih Lee. It is your vision, core values, education and family that enable you and your company to better yourself and the company and make innovations and improvements for the benefit of patients as shown in Fig. 19.1.

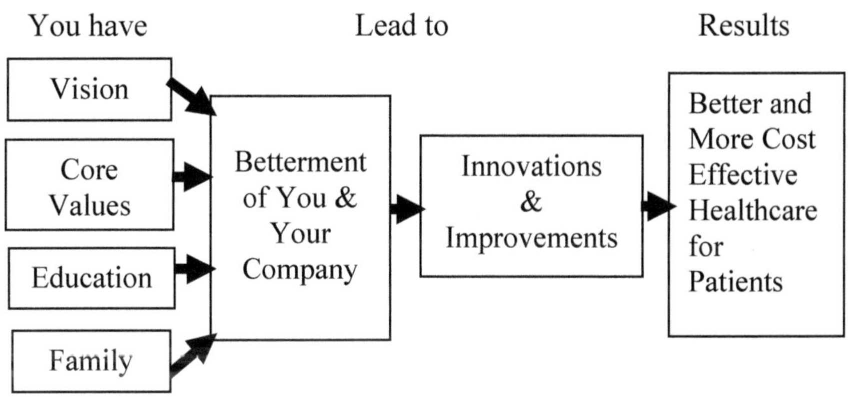

Fig. 19.1. Some key factors for you and your company in creating innovations and improvements that benefit patients.

19.2. Contributing to the Profession, the Society and the Nation

On BME Education, Michael Khoo, Ph.D., Professor and D.C. and H.E. Baum Chair, Biomedical Engineering Department, University of Southern California

Biomedical engineers, who have become successful entrepreneurs, are also fountains of knowledge that can be tapped to relate to subsequent generations their experience and how they managed to convert engineering ideas into devices, products or methodologies that can be applied in the clinical arena. The mechanisms of technology transfer, the complicated web of medical product regulation, and the economics of medical device development — these issues that need to be confronted head-on by biomedical entrepreneurs, which are not generally addressed in the biomedical engineering curriculum. Successful biomedical entrepreneurs can play a critical role as mentors and role models in the education of up-and-coming biomedical engineers, who plan to pursue careers in industry.

On Biomedical Engineering as a Natural Home of Entrepreneurship, Peter G. Katona, Sc.D., President, The Whitaker Foundation

Mr. U.A. Whitaker was a successful engineer, entrepreneur, and businessman whose testament created a foundation in 1976 for "charitable, scientific, literary, and educational" purposes. Since Mr. Whitaker had believed that engineering could make major contributions to medicine, The Whitaker Foundation dedicated itself to improving health by enhancing the field of biomedical engineering. Through a variety of grant programs, the foundation helped to establish the research careers of more than 1,000 faculty members, and to triple the number of biomedical engineering departments, which now exceeds 80. Such an investment was motivated by the belief that educating a new generation of biomedical engineers would bring about the strongest and longest lasting benefits (7).

Engineers are often considered to be professionals who provide technological solutions to well specified problems. In the life sciences and medicine, however, the problems are often ill defined, and the major task is asking the right questions when seeking answers. The hallmark of a well-educated biomedical engineer is that he or she can participate substantively in identifying important problems, in addition to contributing to technological solutions. This is achieved by drawing upon a rigorous integrated training in both the life sciences and engineering.

Identifying problems is especially important for entrepreneurs. These individuals make their contributions by possessing an in-depth understanding of needs and coming up with solutions that are effective, novel, and — hopefully — profitable. They do not design devices that have been specified by others. Instead, they design devices, materials and processes that solve real biomedical problems that they recognize and formulate.

Before its planned closing in 2006, The Whitaker Foundation conducted a survey of the recipients of its biomedical engineering research grants. Almost half of the 614 responders had at least one patent, and more than a third created new products, resulting in more than 100 new companies (7). This interest in entrepreneurship is especially noteworthy since the research grants had been designed for faculty members early in their careers, rather than for practicing engineers.

Biomedical engineering is a profession that requires knowledge in diverse areas, welcomes risk, poses major challenges and offers substantive rewards. Thus, it is a natural home for entrepreneurship. Mr. Whitaker would be pleased.

On Commitment to Invention, Collaboration, and Mentorship (The September 19, 2017 announcement on awarding the Lemeson MIT Award to **Feng Zhang**[a], W.M. Keck Center Development Professor of Biomedical Engineering, the James and Patricia Poitras Professor in Neuroscience at MIT, and associate professor in the departments of Brain and Cognitive Sciences and Biological Engineering at MIT)

Zhang and his team were first to develop and demonstrate successful methods for using an engineered CRISPR-Cas9 system to edit genomes in living mouse and human cells and have turned CRISPR technology into a practical and shareable collection of tools for robust gene editing and epigenomic manipulation. CRISPR, short for Clustered Regularly Interspaced Short Palindromic Repeats, has been harnessed by Zhang and his team as a groundbreaking gene-editing tool that is simple and versatile to use. A key tenet of Zhang's is to encourage further development and research through open sharing of tools and scientific collaboration. Zhang believes that wide use of CRISPR-based tools will further our understanding of biology, allowing scientists to identify genetic differences that contribute to diseases and, eventually, provide the basis for new therapeutic techniques.

Zhang's lab has trained thousands of researchers to use CRISPR technology, and since 2013 he has shared over 40,000 plasmid samples with labs around the world both directly and through the nonprofit Addgene, enabling wide use of his CRISPR tools in their research.

Zhang began working in a gene therapy laboratory at the age of 16 and has played key roles in the development of multiple technologies. Prior to harnessing CRISPR-Cas9, Zhang engineered microbial TAL effectors (TALEs) for use in mammalian cells, working with colleagues at Harvard University, authoring multiple publications on the subject and

[a] Zhang delivered the NIH NIBIB Lecture: From Microbial Immunity to Genome Editing at the BMES 2017 Annual Meeting in Phoenix.

becoming a co-inventor on several patents on TALE-based technologies. Zhang was also a key member of the team at Stanford University that harnessed microbial opsins for developing optogenetics, which uses light signals and light-sensitive proteins to monitor and control activity in brain cells. This technology can help scientists understand how cells in the brain affect mental and neurological illnesses. Zhang has co-authored multiple publications on optogenetics and is a co-inventor on several patents related to this technology.

Zhang's numerous scientific discoveries and inventions, as well as his commitment to mentorship and collaboration, earned him the Lemelson-MIT Prize, which honors outstanding mid-career inventors who improve the world through technological invention and demonstrate a commitment to mentorship in science, technology, engineering and mathematics (STEM).

"Feng's creativity and dedication to problem-solving impressed us," says Stephanie Couch, executive director of the Lemelson-MIT Program. "Beyond the breadth of his own accomplishments, Feng and his lab have also helped thousands of scientists across the world access the new technology to advance their own scientific discoveries."

"It is a tremendous honor to receive the Lemelson-MIT Prize and to join the company of so many incredibly impactful inventors who have won this prize in years past," says Zhang. "Invention has always been a part of my life; I think about new problems every day and work to solve them creatively. This prize is a testament to the passionate work of my team and the support of my family, teachers, colleagues and counterparts around the world."

The $500,000 prize, which bears no restrictions in how it can be used, is made possible through the support of The Lemelson Foundation, the world's leading funder of invention in service of social and economic change.

"We are thrilled to honor Dr. Zhang, who we commend for his advancements in genetics, and more importantly, his willingness to share his discoveries to advance the work of others around the world," says Dorothy Lemelson, chair of The Lemelson Foundation. "Zhang's work is inspiring a new generation of inventors to tackle the biggest problems of our time."

On Demand for Healthcare, Dane A. Miller, Ph.D., President and CEO, Biomet, Inc. (in the Whitaker Foundation 2005 Annual Report)

There are three basic techniques for dealing with the increased demand for healthcare in the next quarter century. One option would be to reduce compensation provided to those delivering the healthcare. The second choice would be to deliver fewer units of healthcare by rationing. The third possibility would be to create advances in technology so that more efficient and higher quality healthcare can be delivered at a lower cost. It is this third technique that offers the greatest challenge and opportunity for the field of biomedical engineering. Approach no. 1 will simply not work because eventually the healthcare system will drive away suppliers and providers. Approach no. 2 involves a number of socially unacceptable decisions. Therefore society's focus should be on no. 3.

On Drug Pricing and Drug Innovation (Jon Northrup, CEO, Horizon Biotechnologies, 8).

Today the USA is the largest and most diverse market for healthcare, in solutions, jobs, and spending. It has acted as a magnet to healthcare researchers, physicians, and companies, and has helped to create a more concentrated and productive pharmaceutical sector that is the envy of the world. Europe, and now Japan, has moved away from this model to a more controlled, lower priced, and potentially even innovation-hostile pharmaceutical marketplace.

The difference in price for pharmaceuticals between the US and some other countries has created a large political question for the US. Price differences arise from exchange rates, different starting prices, price freezes, and, in some cases, the threat of compulsory licensing (essentially seizing the company's intellectual property) to achieve a low purchase price.

Pharmaceutical research and development is complex, time-consuming, and highly risky. Yet with it comes the potential of a better and longer life for all those who can participate in the fruits of the industry. The challenge ahead is how to routinely and efficiently create drug innovation so that these benefits can be delivered at a price that the world can afford. Some hope exists in the biological revolution presently underway and in the potential to understand disease and its biological roots. Some hope exists in informed consumers who understand the value of the intangible effort that goes into the creation of a small pill with unique properties.

On Patients as Partners, Dennis Bourque, Founder, Executive Director and Mentor, Renal Empowered Mentors for Education in Nephrology & Dialysis (ReMend) Organization

Much has been discussed on how innovative technologies can improve patients' lives. However, technology by itself can be limited in improving patient outcomes. Despite being a technology executive for 30 years, as a lifelong chronic kidney disease (CKD) patient and an active member in the medical community, I have met hundreds of patients who are in need of significant technology related help.

When most CKD patients are first diagnosed, they are told by their doctor to begin in-center dialysis right away. However, most patients don't have the knowledge to understand the incredible impact that dialysis will have on their lives and are not sufficiently educated to make the difficult decisions related dialysis treatments. Thus, most patients resort to the most frequently used option, which includes receiving dialysis treatments in a clinic three times a week at three to four hours per visit. This naturally becomes an incredible disruption to their quality of life and mental attitude and significantly impacts their work schedule if they are able to continue to work.

However, there is a much less disruptive form of dialysis that can be received at home without a major disruption in the patient's life, has better patient outcomes and costs less. Although this home-based dialysis technology has been available for decades, only 10% of CKD patients choose this option. One of the reasons for not seeing a higher utilization of newer medical devices is that the patient's medical team lacks the basic training and understanding of the breadth of benefits that range from how new and complex technologies can help to combat the patient's disease to how they impact the patient's experience. These are key factors in how patients make these difficult, life-changing decisions.

At ReMend, our mentors are long-term CKD patients themselves. Thus, they can become active partners with the medical team and share their patient coping and resiliency skills so that everyone becomes better educated and patients become empowered by their knowledge of their disease. As today's technology is often overly complicated, it becomes disruptive to educate both the patient and medical team in it. This can even be a barrier for the patient and their caregivers as they attempt to become better educated and knowledgeable of their disease.

The next logical question is what can be done to change these technology barriers. To begin, we should have patient engagement, which is taking the time to truly understand the patient's perspective. How patients consider these difficult decisions also provides an opportunity to tailor patient education and address fears and concerns head on. At ReMend, this is what we do. We leverage our CKD patients as mentors who have already been through these difficult decisions. The mentors provide firsthand patient experience and education that benefits new patients in making sound decisions for themselves. Similarly, exposing the benefits of knowledgeable patient feedback and engagement provides medical professionals and technologists opportunities to improve patient care that was overlooked in the past.

Please take the time to partner with patients, caregivers and their medical teams, especially to solicit their input when you develop the life-changing technologies that will forever change patients' lives.

On Partnerships, Harold Varmus, M.D., Director of National Institutes of Health (1993-1999), Foreword, 1998 NIH Symposium on Bioengineering: Building the Future of Biology and Medicine

Bioengineering advances the nation's health by applying engineering principles and techniques to biological problems. The rewards that are most obvious to the public are novel devices and drugs. However, but bioengineering also offers further insight into biological processes, new methods for using data from genetics and an increased ability to visualize the brain and other organs. History tells us that most of the revolutionary changes that have occurred in biology and medicine have depended on new methods that are, themselves, often the result of fundamental discoveries in many different fields. Thus biological problems are too complex to solve by biologists alone. We need partners in many disciplines, including physics, mathematics, chemistry, computer sciences and engineering.

On US Healthcare, Karen Davis, President of the Commonwealth Fund, New York (4)

High healthcare expenditures and the growing number of people without health insurance set the United States apart from all other industrialized countries. The United States spends twice per capita what other major industrialized countries spend on healthcare, but is the only one that fails to provide near-universal health insurance coverage. We also fail to achieve health outcomes as good, or value for health spending as highly, as are achieved in other countries.

The United States has been slow to learn from countries that have systematically adopted policies that curtail spending and enhance value. Chief among these are mechanisms for assessing the comparative cost-effectiveness of drugs, devices, diagnostic tests, and treatment procedures and implementation of information technology. These latter includes electronic repositories of patient medical information across sites of care: easy access to primary care, including organized systems of off-hours care; a strong role for government in negotiating payment for care and payment systems that reward preventive care, management of chronic conditions, care coordination, and health outcomes rather than volume of services.

On Pricing Your Products, Money Magazine, December 2016

Never underestimate the power of parents who feel their children's lives are endangered because of corporate greed.

Outraged parents unleashed their fury on pharmaceutical maker Mylan for "unconscionable price gouging" of a product that their children literally could die without — leading to a congressional hearing, $500 million in penalties and savings for thousands of families.

Mylan owns the EpiPen, a device that wards off life threatening anaphylactic shock during allergic reactions. Roughly two children in every U.S. classroom have allergies that warrant an EpiPen prescription. Families tend to restock EpiPens, which expire annually, during back-to-school season. It was then that many realized that their high-deductible insurance plans left them on the hook for the full retail price --- over $600 per pack, up from less than $100 when Mylan acquired the product in 2007. Families need multiple packs (one each for school, home, and to carry around). Consequently, bills hit the thousands.

Parents reacted by flooding Mylan's Facebook page with comments that assailed the price increases. They wrote to politicians to demand action. This led to a congressional hearing and public condemnations of Mylan.... Hundreds of thousands signed online petitions to plead for sensible pricing, and allergy advocacy blogs were filled with heartbreaking personal stories.

“Parents were saying, ‘What am I supposed to do, not pay the mortgage? Not pay for groceries?'" says AllergyKids.com founder

Robyn O'Brien. "I heard from school nurses, doctors, pharmacists, members of Congress--everyone was enraged."

Months of this bad publicity prodded Mylan into releasing coupons for bigger EpiPen discounts and launching a cheaper generic version. The uproar also led federal investigations to reveal that Mylan had overcharged the Pentagon and Medicare for EpiPens to the tune of $500 million.

Perhaps most important, the EpiPen saga brought to light one of the most dysfunctional parts of our health care system. "Hopefully, this kicks off a much bigger conversation about how we price medicine." says O'Brien.

On the Biological Revolution, Robert Nerem, Ph.D., Parker H. Petit Professor and Director of the Institute for Bioengineering and Bioscience, Georgia Tech

This 21st century is being called the "Biotech Century." It is called this because the biological revolution, which arguably started at the beginning of the 20th century, has advanced so significantly that it provides the foundation for the new technologies of this 21st century. Advances in biology are continuing to advance at an ever accelerating pace and, in addition, the world of technology has become global in nature. These developments have several implications.

First, the medical device/implant industry already is being impacted by the convergence of biology and the traditional electro-mechanical technologies that were developed in the second half of the last century. As important as these have been for healthcare, the emerging combination products (i.e., the combination of a biologic and a device or a drug and a device) offer new possibilities for treating patients. Also emerging are regenerative medicine and the use of stem cells of various types in the replacement, regeneration and/or the repair of tissues and organs. This will provide the opportunity to develop new therapies that are cell-based and will facilitate the treatment of diseases and conditions that currently have none.

Second, this biological revolution has impacted, and continues to do so, not only the medical arena, but all of biology. This means that biomedical engineering must evolve into a much broader biology-based engineering, which more appropriately should be called bioengineering. This is because one might predict that, in 15 to 20 years, the application and impact of biology-based information will be just as great outside of the field of medicine as within it. This new field of bioengineering already is beginning to address major concerns of society, such as energy from biofuels, the production of food by agricultural biotech, and the environment. Undoubtedly, will be more areas to come.

Third, the global nature of our society demands that a biology-based engineering not only address problems of the developed countries, but also those of the developing countries. In Africa and parts of Asia there is a need for healthcare technologies that not only can treat the diseases prevalent there, but also are cost effective and easily delivered. In these developing countries, food is a major issue and, increasingly, there are environmental issues.

Thus, the application of a biology-based engineering (i.e., bioengineering) is continuing to broaden and diversify. With this, the need for new biology-based technologies and products will also continue to broaden, which will require a continuing emergence and broadening of the field. Although it is always difficult to predict the future, it is clear that the future is bright. Despite all of the changes that have occurred in the last 30 years, these driven by the biological revolution and the investments made by The Whitaker Foundation, there will be even more profound changes in the next 30 years. Thus, the future of this field of biologically-based engineering is exceedingly bright, although it is very likely that it will be much different than what any of us envision today.

On Public Policy and Innovations, Duane Roth, Chief Executive Officer, in the 2008 Annual Report of CONNECT, San Diego, California

To prepare for the next 25 years in an increasingly competitive global innovation economy, CONNECT has adopted the following key objectives that are fundamental to our future competitiveness:

- Research institutions are the engines of discovery and we should unite to recruit and retain the most talented researchers anywhere and offer support for increased federal funding of basic research;
- Intellectual property is our primary asset and we should be diligent in protecting our discoveries around the world;
- The global economy requires ever increasing partnerships to remain competitive and, as a region, we should reach out nationally and internationally and share our contacts;

- Communication between academic leaders and the private sector will be even more essential to the innovation economy and we should develop ways to continue to improve;
- Risk capital is the fuel of innovation and we must develop new and sustainable models to fund innovation; and
- Recruitment of talented entrepreneurial leaders and creative scientists and engineers is essential and we need to develop tools to ensure that we remain competitive.

If one evaluates these objectives, it becomes apparent that public policy in Washington D. C. and Sacramento holds many of the keys to our future success. Federal support of research is the lifeblood of new idea generation. Intellectual property laws, trade policy and regulation, set largely by Washington policy makers, will determine much of our global competitiveness. As a region, we must become more engaged with our legislators and CONNECT is committed to improving communication with our Washington and Sacramento legislators.

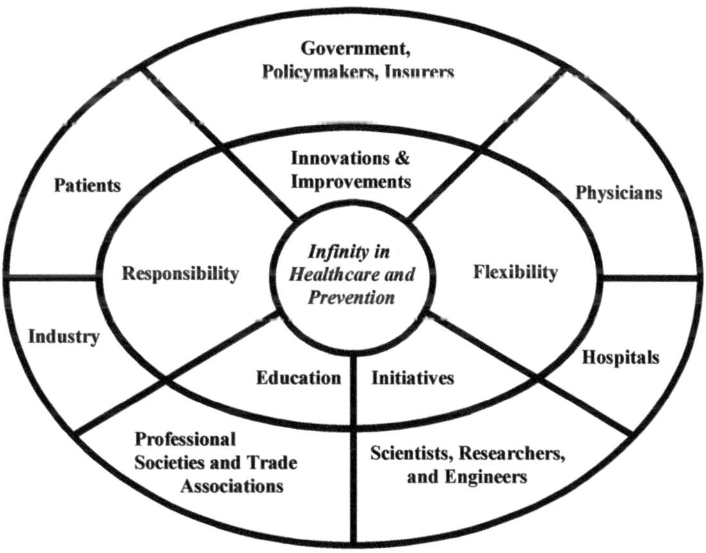

Fig. 19.2. Healthcare and disease prevention for the improvement of people's health are unlimited when their constituents (the outer ring) work together through education, responsibility, flexibility, initiatives, improvements and innovation.

Conclusion. Jen-Shih Lee. The challenges and opportunities in healthcare and disease prevention are infinite. To deliver better healthcare at lower cost and to be more effective in prevention, the constituents that are identified in the outer ring of Fig. 19.2 need to work together responsibly, educate themselves, show flexibility, take the initiative and create innovations and improvements.

19.3. Doing Something for the World

On Global Health. Bill Gates, the James C. Morgan Global Humanitarian Award delivered to the Tech Museum, 2006 (9).

For a long time, I continued to see my largest contribution to the community in terms of creating a successful business. I just didn't have the time and energy to do more.

And then one day, Melinda and I read an article about the millions of children who were dying every year in poor countries from diseases that we had long ago eliminated in this country. One disease I had never even heard of — rotavirus — was killing literally half a million kids each year. I thought, "That's got to be a typo. If a single disease was killing that many kids, we would have heard about it — it'd be front-page news." But it wasn't.

We couldn't escape the brutal conclusion that — in our world today — some lives are seen as worth saving and others are not. We said to ourselves, "This can't be true. But if it is true, it deserves to be the priority of our giving."

That became the core principle of the Gates Foundation: every human life has equal worth. It seemed clear to us that the most intractable problems and the greatest suffering came as a result of inequality — in access to education and economic opportunity, across race, countries, and continents.

The death of a child in a poor country is every bit as tragic as the death of a child in a rich country, and those of us who were born in rich countries have a responsibility to help those who weren't. We believe public health is a lever for lifting people's lives.

When health improves, life improves, by every measure — from higher literacy, to better education, to stronger economic growth and a more stable, prosperous society.

And we believe that technology can be used to advance public health in every way — from discovery, to development, to delivery.

On Global Health by Design. John Linehan, Ph.D., Professor of Biomedical Engineering and Medicine, Northwestern University

Bill Gates is correct. Technology can be used to advance public health — if it is affordable! Global Health by Design is a great way to involve students. They readily respond to the challenge of designing devices to address under-met clinical needs in resource-limited settings. The key to success will be to minimize cost by using existing technology in ways that meet the needs of the end user and any limitations that are imposed by their environment.

Northwestern University provides an example. Four years ago, NU established a medical-device innovation program, Global Health Technologies (GHT), in collaboration with the University of Cape Town (UCT) in South Africa. Four devices that have been developed with essential student participation within the GHT program are either in, or will soon be in, the clinical testing phase of device development.

These device-design projects were initiated in collaboration with the UCT faculty, and in close coordination with healthcare professionals and medical administrators of the Western Cape, and African businesses in the medical and non-medical technology sectors. The geographical reach has gradually broadened to include participants in other South African provinces and other countries in southern African.

The GHT program focuses on device design and innovation, taking a need-based and user-centered approach to solving problems that are specific to under-resourced and medically over-burdened regions of sub-Saharan Africa. The program is under the purview of both the Biomedical Engineering Department and the NU Center for Innovation in Global Healthcare Technologies (www.cight.northwestern.edu).

Student who have participated include undergraduates, MS students and PhD students from both Northwestern University and the University of Cape Town.

On The Smile Train. Charles B. Wang, CEO and Chairman, Computer Associates and co-Founder and Chairman, The Smile Train (10)

Wang said today (March 14, 2009) that The Smile Train, an international children's charity and medical organization, is off to a great start following an international symposium where $20 million was donated to fund a series of joint Sino/U.S. strategic initiatives to help eradicate cleft lips and palates.

"The Smile Train has brought together the finest doctors, nurses, surgeons and scientists from China, the United States and around the world," said Bush (41st President of USA). "Although they may speak different languages and come from very different backgrounds, there are some things they all have in common: the belief that every child counts, and that every child deserves a smile."

"Our initial commitment of $20 million is just the beginning of what it will cost to help the millions of children with clefts in China who cry themselves to sleep every night," said Wang, Founding Member of The Smile Train. "Just imagine what it must be like to live your entire life without being able to smile. That's why we need so many more people to get on board The Smile Train and help us help these children."

Unlike other organizations that rely on American surgeons to perform surgery in foreign countries, The Smile Train believes that local self-sufficiency is the only realistic, cost-effective answer to the problem. Through free education, training and equipment, local medical professionals and teams can be empowered to provide the help that is needed by the local children.

"I can think of nothing more special and worthwhile than helping disadvantaged children who have no place else to turn," said Wang. "And there's no organization that I know of that has the potential to change the lives of as many children as The Smile Train." It can take as little as 45 minutes for cleft lip surgery and up to three hours for cleft palate surgery. "Smile Train surgery is the closest thing to a modern-day miracle that I have ever witnessed," said Wang. "It's like giving a child a second chance in life."

In 1999, its first year of operation, The Smile Train funded programs and partnerships at a level of $10 million to $15 million. Since Wang's private contributions have covered all administrative costs of The Smile Train, 100% of all additional donations will go to program areas of training, treatment and research.

(In the ten years since the founding of The Smile Train by Wang and Brian Mullaney, The Smile Train has provided surgery to 500,000 children. Smile Pinki in Fig. 19.3 is one of these children. With a cost as little as $250 and a rate of 100,000+ surgeries a year now, 500,000 desperate children not only have just a new smile, but also a new life.)

Fig. 19.3. A joint effort to solving global health problems of the third world (courtesy of The Smile Train with Pinki's picture, 10).

Conclusion. Jen-Shih Lee. Global health and world economy are the essentials for all of us to **collaborate, share and work on** in this 21st Century as shown in Fig. 19.4 of the next page. The three elements — affordable technology, local effort and running philanthropy like a business as depicted in Fig. 19.3 above are bound to deliver not only smiles to many children like Pinki, but also better healthcare and economy for all people in the world.

Fig. 19.4 Collaborate, share and work for the people and the world.

References

1. Bakken, E., *Reflections on Leadership*, Medtronic, MN (2001).
2. Chang, CM, *Engineering Management, Challenges in the New Millennium*, Pearson Prentice Hall, N. J. (2005).
3. Collins, JC and WC Lazier, *Beyond Entrepreneurship*, Prentice Hall, N.J. (1992).
4. Davis, K., *N. Engl. J. Med.*, 359: 1751-1755 (2008).
5. Hawkins, B., *MD&DI*, p.32 June (2009).
6. Jarvik, R., Time, p.18, March 2 (2009).
7. Katona, P.G., Biomedical Engineering and The Whitaker Foundation: A Thirty-Year Partnership. *Annals Biomed. Eng.* 34: 904-916 (2006).
8. Northrup J, The pharmaceutical sector, in *The Business of Healthcare Innovation*, ed. L.R. Burns, Cambridge Univ. Press; pp.27-102, (2005).
9. www.microsoft.com.nsatc.net/presspass/exec/billg/speeches/2006/11-15TechMuseum.mspx
10. www.smiletrain.org

Chapter 20

How to Succeed in Biomedical Engineering Entrepreneurism with Really Trying*

Shu Chien, M.D.
Professor of Bioengineering and Medicine, UCSD
Director, Whitaker Institute for Biomedical Engineering, UCSD
President, Biomedical Engineering Society (2006-2008)

Dr. Jen-shih Lee has written an excellent book for undergraduate and graduate students, as well as working biomedical engineers, who want to learn the basic knowledge required for the development of biomedical engineering entrepreneurship. In the Chapter on *Investing in Biomedical Engineering*, he has presented the results of his interviews with many successful biomedical engineering entrepreneurs and visionary educators in the field in three Sections. The superb visions, insightful analysis, and valuable advices by the interviewees make this an extremely valuable Chapter. It is a privilege for me to be asked by Dr. Lee to write an excerpt. I have made an attempt to summarize these marvelous paragraphs without referring specifically to the Sections, because there are some interplays between them. Because many of the important thoughts were expressed by several interviewees and for the sake of continuity in reading, I have not made specific references to them in this excerpt. In a few places I have added my own thoughts.

20.1. The Biomedical Engineer as a Person

Core Values. The biomedical engineers, including those who become biomedical engineering entrepreneurs, must possess and/or develop

---

* The title of this chapter is a take-off from a comedy movie made in 1967 entitled "How to Succeed in Business without Really Trying?", starring David Morse, Maureen Arthur, and Rudee Vallee.

certain personality traits, many of which are common for scientists and engineers, as well as other citizens of the Society. The biomedical engineer must have the core values of integrity, honesty, dedication to truth, fairness, passion, generosity, discipline, perseverance, courage, sense of responsibility, respect, and loyalty. A very important element to be successful in entrepreneurism, as well as biomedical engineering in general, is innovation.

Balance in Life. Biomedical engineering entrepreneurship is serious business that takes a lot of time, energy and concentration. It is important, however, to balance the many priorities in life. This includes personal health (physical and mental) and family life. Therefore, time management is very important. A healthy mind and body and a happy family can have major contributions to success in biomedical engineering entrepreneurship, just as in other professions. Work, health, and family are interdependent. The key is how to manage time to enjoy all these elements and keep them in proper perspectives.

20.2. Development of a Biomedical Engineering Entrepreneur

In order to develop into a successful biomedical engineering entrepreneur, it is important to have an excellent background and education, which should integrate natural sciences, life sciences and engineering. The training should foster the ability to identify important problems and the approaches to solve them, as well as the development of a vision for the future. One needs to learn how to convert ideas into products (the work "product" will be used to include devices, technologies, biologics, drugs, etc.) that can be used in clinical medicine. By knowing what is the existing state-of-the-art in terms of opportunities and challenges, the entrepreneur will be able to integrate these into his/her thought process to develop innovative ideas and pursue them. In order to be innovative, one needs to think differently, boldly, creatively, and unconventionally.

Besides the formal training in schools, the influences of outstanding mentors or role models have often been cited as an important element in career successes. Such mentors provide the role models for being innovative, visionary, dedicated, and all the other essential characters for success in biomedical engineering entrepreneurism. It is essential that the young biomedical engineers being trained today learn how to mentor and educate the next generation, so that the development of biomedical engineering and entrepreneurism can be sustained and amplified.

20.3. Inter-personal Interactions

No person can live as an island. The need for interactions with other people is particularly important for biomedical engineering entrepreneurs, who work in a team environment in the generation and delivery of products. The importance of the relations with family members and mentors has already been mentioned. Engineers with excellent managerial skills and superior business acumen, in addition to their superb technical capabilities, are needed to lead Corporate America to turn technological innovations into profitability. A study by the National Science Foundation estimated that almost one half of engineers or scientists will be engaged in managing people, projects, teams, technology and other resources to add value to their companies.

In industrial companies, the biomedical entrepreneur works with other engineers, the management, technical and supporting staff, and field representatives. In delivering the products, he/she works with physicians, other health professionals, and the patients (though mostly indirectly). Positive and effective interactions with all of them are critical. In the design and manufacture of products, and in marketing and advertising, it is essential to have the spirit of "customer first". In the final analysis, the aim of these products is to improve the health of the patient and treat their illness. The positive outcome in patients drives the passion of the entrepreneur, the productivity of the company, and the success of the business. It is always the patients first.

20.4. Roles of the Government and the Private Sector

The funding support by the National Institutes of Health (NIH), National Science Foundation, and the other government agencies for research and training has played an important role in fostering the creation of novel ideas, development of new products, and education of the next generation of biomedical engineers. The Small Business Innovation Research (SBIR) and Small Business Technology Transfer (STTR) programs are especially valuable in fostering the establishment and development of small business, with a significant number related to biomedical engineering.

The unprecedented, strong, and dedicated support of biomedical engineering by The Whitaker Foundation has made extraodinary impacts on the development of the discipline from 1976 to 2006. The Foundation helped to establish the research careers of thousands of faculty members, and to triple the number of biomedical engineering departments. The fostering of the education of a new generation of biomedical engineers has brought about strong and long-term benefits. They have not only excelled in the science of biomedical engineering, but also contributed importantly to entrepreneurism, through the design of devices, materials and processes that solve biomedical problems, filing of over 300 patents, and making possible the formation of over 100 new companies.

More recently, the Wallace H. Coulter Foundation's President Sue Van initiated the Translation Research Partnership Awards to promote translational research by enhancing collaborations between biomedical engineers and clinicians and utilizing best practices of industry to accelerate academic innovations to the market to improve patient care. The Awards support the movement of promising technologies to clinical applications and develop sustainable processes for science to serve humanity.

The Thomas Siebel Scholars Foundation has extended its Siebel Scholars Program to Biomedical Engineering) Ph.D. students who have outstanding academic achievements, a track record of distinguished research, and excellence in leadership qualities. Siebel Scholars form a community of leaders and are key advisors to the Siebel Foundation in guiding the development of innovative programs, including science, public policy and entrepreneurism.

The Howard Hughes Medical Research Institute (HHMI) and other foundations and institutes have also begun to support biomedical engineering education, including translation to clinical medicine and industrial products. Professional Societies such as the Biomedical Engineering Society (BMES), IEEE/Engineering in Medicine and Biology, and many others have been playing significant roles in fostering entrepreneurism for students and other members.

While the funding for research and education in the United States is at the top of the world, its rate of change is rather alarming. The annual increase in NIH funding is barely in keeping with inflation. The American Recovery and Reinvestment Act (ARRA) in FY 2009 and 2010 provided a one-time infusion of $10.4 billion to an existing annual

NIH budget of less than $30 billion. Because of the short lead time, it was not possible to have comprehensive planning for this large amount of incremental funding. Because of the short duration of implementation, it was not possible to have continuity and sustenance. Such pulse injection of funding could not achieve the purpose of sustained enhancement of biomedical research, which can only be attained by steady, planned increases in funding. In comparison to some other countries, e.g., China, the rate of % change of U.S. funding support is markedly behind. There is an urgent need to enhance funding support for research and training from the government and the private sector in order to sustain and enhance the engines of discovery in universities and research institutions.

20.5. Public Policy and Healthcare Delivery

The convergence of the remarkable advances in biology with the rapid developments in novel engineering technologies has impacted healthcare, e.g., the emergence of combination products (e.g., biologic and device or drug and device) offer new therapeutic possibilities. The use of various types of stem cells in regenerative medicine provides the opportunity to develop cell-based and other novel therapies for the treatment of diseases heretofore not possible. As new therapeutic and diagnostic procedures become increasingly available, there is the important question of rising healthcare expenditures; therefore, it is important to develop policies to assess comparative cost-effectiveness in order to curtail spending and enhance value. This is particularly important in the US because of the growing number of people without health insurance.

With the rapid advances in healthcare approaches, it is important that our regulatory procedures can adapt to these increasing demands by making the best cost-effective therapeutic modalities available to the public while protecting them from potential risks of untoward effects.

Intellectual property is the primary asset for the innovative entrepreneurs, and we should be diligent in maximizing the effectiveness for protecting their discoveries. Risk capital is the fuel of innovation, and we must develop new and sustainable models to fund innovation.

The important role of federal funding in fueling and translating novel discovery has already been mentioned. Organizations such as the Federation of American Societies for Experimental Biology (FASEB),

Research!America, American Medical Association, AIMBE and others have been active in providing inputs for federal funding and other public policy matters. It is important that individual biomedical engineers join in these efforts to enhance our voice from the biomedical engineering community and our communication with our legislators and our society.

20.6. Global Nature of Biomedical Engineering Entrepreneurism

With the rapid development of wireless communication, science and engineering have become increasingly global with vanishing national boundaries, and this is also the case for biomedical engineering entrepreneurism. Global economy also requires enhanced partnership to reach out nationally and internationally and share our contacts. We need to interact with and learn from each other on the various matters mentioned above. An example is the adoption of policies that curtail spending and enhance value, e.g., assessment of comparative cost-effectiveness of therapeutic procedures, implementation of information technology, easy access to primary care, role of government in negotiating payment for care, and payment systems that reward preventive care, management of chronic conditions, care coordination, and health outcomes.

As a result of the global nature of our society, biomedical engineering must address health problems in not only the developed, but also the developing and underdeveloped, countries. We must view every human life as equal in worth. Inequality in access to education and economic opportunities by disadvantaged people across the world leads to intractable health problems and inequality in healthcare. We need to use public health to improve people's health and lives. We should develop healthcare technologies that are not only effective, but also cost-effective and easy to deliver. Public health can be advanced by using biomedical technologies from discovery to development, and to therapeutical delivery.

20.7. Conclusions

In order to be a successful biomedical engineering entrepreneur, the individual must have the basic core values and characters, including passion and dedication for the work, but these must be balanced with other priorities in life: health and family.

The development of a biomedical engineer entrepreneur requires an excellent interdisciplinary background and education, with the abilities to identify important problems and to convert ideas into innovative products for clinical application. It is important to have outstanding mentors or role models in career development.

Biomedical engineering entrepreneurs need to interact with other engineers, the management, technical and supporting staff, field representatives, physicians, other health professionals, and patients, with the final aim of improving the health of patients and treat their illness. Managerial and inter-personal skills are particularly important for those who would take a managerial position.

The funding support by government agencies for research and training has played an important role in fostering the creation of novel ideas, development of new products, and education of biomedical engineers. The unprecedented, strong, and dedicated support of biomedical engineering by The Whitaker Foundation made a major impact on the development of the discipline. There is an urgent need to increase research and training funding by governmental agencies and foundations to sustain and enhance the fountain of innovation and engine of discovery.

The convergence of the remarkable advances in biology with the rapid developments in novel engineering technologies has impacted healthcare, and there is the possibility of concurrent rise in healthcare expenditures. It is important to develop policies to assess comparative cost-effectiveness and to adapt our regulatory procedures. It is also important to develop models for intellectual property and risk capital. Biomedical engineers should enhance our voice regarding funding and other matters, and increase communications with our legislators and our society.

Biomedical engineering entrepreneurism has become increasingly global. Global economy requires enhanced partnership to reach out and share our contacts. We need to interact with and learn from each other on the various matters mentioned above related to healthcare delivery and health policy. Biomedical engineering must address health problems in developing countries and value every human life as equal in worth.

The various topics discussed above can be summarized in a diagrammatic form in the next page (Fig. 19.1).

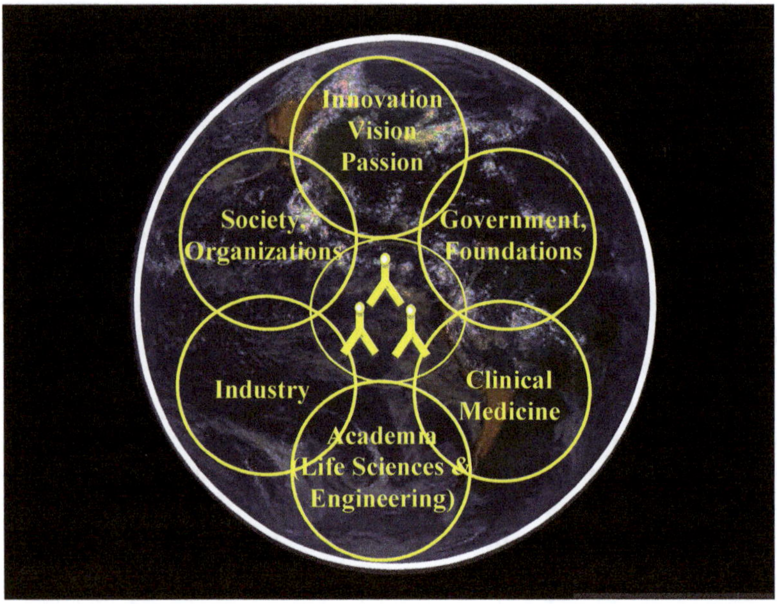

Figure 20.1. A schematic diagram to illustrate the thoughts presented in this Chapter. In the center of the diagram, each of the three lambda-like figures represents symbolically a person: It is the Chinese character for a person 人. Each person must have the core values and be able to balance his/her life, as mentioned in the text. It is essential for the person to have ideas for innovation, a clear vision, and a strong passion for his work and for other people, especially the patients. In the Chinese language, the combination of three such person symbols becomes a group of persons, or people 众 Thus, people are in the center of everything, and it is essential to have interactions among people, including that between entrepreneurs and colleagues, physicians, and patients, that between mentor and mentee, and many others.

The bottom part indicates the importance of having a strong interaction among academia (including life science and engineering), clinical medicine, and industry. The biomedical enterprise needs the support of government and foundation, and it is essential to have effective professional organization and to interaction with the society. The large round background indicates the global nature of these activities and objectives.

Acknowledgments

Many of the passages and ideas were taken from the excellent comments in Chapter 19 of this book by Earl E. Bakken, C. M. Chang, Karen Davis, Bill Gates, Bill Hawkins, R. Jarvik, Peter G. Katona, Michael Khoo, Robert Nerem, Duane Roth, and Harold Varmus. Because several of the thoughts were common to several contributors, the author apologizes that there are no clear identifications of the specific sources.

Index

www.ingramcontent.com/pod-product-compliance
Ingram Content Group UK Ltd.
Pitfield, Milton Keynes, MK11 3LW, UK
UKHW022104170926
8624IPUK00024B/81

9 789813 270428